FIREFIGHTER PROMOTION EXAMINATIONS

—— Practice and Review ——

by

GENE MAHONEY
Fire Science Coordinator
Rio Hondo College

ARCO PUBLISHING, INC.
NEW YORK

Published by Arco Publishing, Inc.
215 Park Avenue South, New York, N.Y. 10003

Copyright © 1983 by Arco Publishing, Inc.

Library of Congress Cataloging in Publication Data

Mahoney, Gene, 1923-
    Firefighter promotion examinations.

    1. Fire fighters—United States—Examinations,
questions, etc. 2. Fire fighters—Salaries, pensions,
etc. I. Title.
TH9157.M25 1983   628.9'2'076   82-18425
ISBN 0-668-05611-8

Printed in the United States of America

CONTENTS

...continued on next page

CONTENTS continued

PREFACE

This book has been developed to assist those preparing for promotion in the fire service. Although the book contains many concepts, principles, and facts useful in establishing a knowledge foundation, it should not be construed to be all that is required in the professional's preparation for promotion. Actually, only the surface has been scratched in most of the disciplines addressed.

With the permission of the publishers, several texts have been used in preparing this book in order to assist students in their study effort. It is recommended that students who wish to further their knowledge refer to these sources for additional information. It is further recommended that those desiring to pursue a particular discipline use the reference material provided in Chapter VIII.

I

HOW TO STUDY FOR PROMOTION

Competing in the examination process can be one of the most rewarding or frustrating experiences in a firefighter's career, depending upon the outcome. Success not only improves a person's position in the department, but also increases the amount of money taken home in the paycheck, fringe benefits, and, ultimately, retirement income. A firefighter who decides to enter the contest should not take the challenge, the competition, or the rewards lightly. He or she should realize that much work is involved, and that success does not occur by accident.

The Competition

Those who compete in fire department written promotional examinations may be divided roughly into three categories:

1. About 50% do little, if any, preparation. These are the people who hope for lucky guesses, and then hope they will do well in other portions of the examination process. These also are the people who generally fail the written examination, and, consequently, miss the opportunity to participate in other portions of the examination process.
2. About 40% make some effort to prepare for the written examination. Some of these people get lucky and get a job. Most, however, end up in the "also-ran" category. These are the people who fail one examination after another, and are just out of the money unless seniority finally gets them over the hump. (In some cases, New York City, for example, this is not even possible.) If these people were to spend as much time preparing for one examination as they spend over a period of years on several examinations, they could probably be successful.
3. About 10% make an all-out effort to be as well-prepared as possible for the written examination. These are the people who are eventually promoted. They may end up in the "also-ran" category on one or two examinations, but eventually they will be successful. This book has been prepared with the assumption that those who will use it fall into this category.

Criteria for Success

Positions on a promotional eligibility list are based on how well candidates do in relation to other candidates, not how well they do according to a standard. It is therefore necessary to make some assumptions regarding the people who compete for these positions:

1. Almost all those hired as firefighters are promotable.
2. Most candidates who make an "all-out" effort to do well on the written examination have similar or nearly equal experience in firefighting and fire protection.
3. Most serious candidates also have similar educational backgrounds.
4. Most candidates devote approximately the same amount of time to preparing for a written examination.

If these assumptions are correct, then success in this competitive field is based more on the method of study than on any other single factor. This being the case, it is extremely important to develop a study method that will produce results. The objective of this chapter is to suggest a method that has proved to be worthwhile.

Factors that must be considered before developing a study method are:

1. What to study.
2. How to gather material.
3. How much time to allocate to various subject areas.
4. How to peak out.

Let's examine each of these individually.

What to Study

When considering what to study, learning to separate the wheat from the chaff is necessary. Information encountered when studying normally falls into one of three general categories:

1. Material which must be learned.
2. Facts that are nice to know.
3. "Who cares" information.

Consequently, it is important to be able to determine the category of study information and to concentrate on material which must be learned and facts that are nice to know. Information in the "who cares" category should be ignored. Surprisingly, however, many people studying for promotion spend substantial time learning material that is useless, and completely ignore information that must be learned. Others assume they know the information that must be learned, and, consequently, devote little of their time to reviewing this material. A system must be developed to prevent these tendencies.

One of the first things to do when attempting to separate material into categories is to evaluate the person writing the examination. In most cases, people who prepare written examinations for firefighter promotional positions are professional personnel employees who are skilled in preparing examinations. However, these people usually have a limited knowledge of fire department organizations and procedures. They rely most on material that appears in books or other publications. IF IT IS NOT WRITTEN DOWN SOMEWHERE, THEY PROBABLY DON'T KNOW IT. Therefore, for study purposes, unpublished information can generally be considered useless.

There are several other facts regarding test preparers. One, they are extremely proud of their profession, and want to prepare professional examinations. Two, they do not like to have their questions protested. Furthermore, if a question is protested, they will try to substantiate the original answer. Questions are generally prepared to withstand protests.

When attempting to separate information into one of the three previously listed categories, the kind of question used in promotional examinations should be considered. Most examinations consist primarily of multiple-choice questions. All the examples given in this book are of that sort. It is therefore important to understand several factors regarding the preparation of multiple-choice questions.

Multiple-choice questions are those that present a problem, establish a situation, or ask a question, and then give the candidate several choices from which to select the best answer. Some information lends itself readily to this kind of question, but it is extremely difficult to prepare questions from other information. Information from which multiple-choice questions can be easily constructed offers logical alternatives. The following does not by any means include all such material, but it should provide candidates with some idea of what is important and should be learned.

<u>Definitions</u>--Definitions form the basis of communication and should always be learned. If a reader does not know the meaning of terms, all concept of what is being discussed is lost. Definitions are also used quite frequently in the preparation of multiple-choice questions. Consider, for example, how easy it is to prepare a question from the definition of ignition temperature:

> The minimum temperature to which a substance in air must be heated in order to initiate, or cause, self-sustained combustion independently of the heating or heated element is best defined as the:

(A) flash point
(B) fire point
(C) ignition temperature
(D) critical temperature

<u>Principles</u>--It is important to consider principles that apply to the profession when assembling study material. As an example, a general principle applicable to supervising people is that all facts should be obtained before disciplinary action is taken. This principle can readily be used in the preparation of a multiple-choice question.

Of the following, the primary obligation of a supervisor in any matter requiring disciplinary action is to:

(A) postpone any action until it can be discussed with the department head.
(B) know what procedures were used in the last similar case.
(C) shift the employee to other work.
(D) learn all the facts of the matter.

Facts--It is extremely easy to develop questions from facts. Take, for example, the following question:

The minimum size main recommended for residential areas is:

(A) 4 inches
(B) 6 inches
(C) 8 inches
(D) 10 inches

Key Words--There are some key words that seem to attract examiners like magnets. Take, for example, the difficulty of protesting a question based upon any of the following statements:

The first true organized efforts toward fire protection began in Rome.
The primary value of the steam expansion is the purging of the air.
Never use water, fog spray, foam, common gas, or liquefied fire extinguishers on magnesium.
Saving life should always be a firefighter's first concern.

How to Gather Material

The first thing that must be done before any attempt is made to gather study material is to determine which books, manuals, etc., will be used as sources. Some examining jurisdictions provide a list of study references; most do not. Therefore, some research is required.

Previous examinations, if available, are the best basis for determining required study material. If possible, the source of each question should be determined as the examinations are analyzed. As mentioned before, a question prepared by a civil service examiner is probably taken from written source material.

If previous examinations are not available, talk to as many people as possible who have competed previously in the examination process for the position sought. Ask each of them to identify the sources of questions used. Also ask for estimates of the percentage of questions from each source.

Analyze examinations from other cities, if available. The examinations for most cities for a particular position are somewhat similar. Arco study manuals also provide some idea of the scope of the questions and the kinds of questions used.

Analyze the scope of the examination as provided in the examination bulletin. It will be necessary to obtain a bulletin from a previous examination. Bulletins usually provide a considerable amount of information regarding the required study material.

Once the reference list of study material has been prepared, it is time to start gathering material. First, read the books and manuals deemed necessary, underlining definitions, principles, facts, and key word statements to be learned.

Underlined information should be extracted from the source book and compiled on 3- by 5-inch index cards. A question should be written on one side of the card and the answer on the other side. The cards should be filed according to source.

It is best to purchase a large box of cards (1,000). Use the box as a filing cabinet. Material set up this way can be revised when changes are made in the source material and thus remain useful for years. New cards can be added and old cards thrown away. For most people, the study process is a continuing effort. This system eases the pain of preparing for future examinations.

Learning the Material

If a good job has been done in gathering the material, it should not be necessary to again refer to the sources. Everything that has to be learned should be on cards. Of course, if a new edition of a book is published, it will be necessary to review it to make sure the information on the cards is still valid, and that new material is added.

Once the cards have been prepared, they can be used for learning the material. Read the question, then try to answer it. If you know the answer, the card can be set aside. If not, the material must be reviewed until the answer is memorized. There are several advantages to this method of study:

1. Cards can be studied on a bus, while waiting for an appointment, etc. A few cards can be carried in a shirt pocket, and reviewed and reviewed until the information is learned.
2. Positive feedback is provided as to whether or not the information has been learned.
3. The cards are easy to revise if changes occur.
4. Separating learned information from unlearned information as study proceeds is a valuable time-saver. It wastes time to review material that is already mastered.

Allocating Time to the Various Areas of Subject Matter

In addition to separating required material from that which is useless, it is vital to allocate study time according to the importance of the material. As an example, it is foolish to devote 50% of the available study time to material that only makes up 5% of the examination. Time should be allocated according to the importance of the material. Analyze the scope of the examination and allocate time accordingly. As an example, suppose questions in an examination are divided as follows:

Rules and Regulations	25%
Supervision Principles	22%
Firefighting	22%
Fire Code	10%
Hydraulics and Apparatus	7%
Arson	4%
Hazardous Materials	5%
Miscellaneous	5%

It would follow that at least 80% of the study time should be allocated to the cards on Rules and Regulations, Supervision Principles, Fire Fighting and the Fire Code; 20% should be devoted to the other subjects. Of course, all material is important and should be learned; however, when time is limited, priority should be given to those areas that count the most on the examination.

Every effort should be made to learn as much as possible during the time devoted to study. Two hours of study are not necessarily two hours of learning. It is possible to reach a point of diminishing returns during a study period. This usually occurs when a person tries to concentrate for too long a period of time. Most adults cannot achieve maximum concentration for longer than fifty minutes. Consequently, it is best to study for fifty minutes, take a break, then resume study for another fifty-minute period. This has been found to produce better results than long periods of concentration.

Peaking Out

Peaking out concerns timing the study process to the examination date. It requires planning so that adequate time is allocated to each area of subject matter. To properly peak, it is necessary that plans start with the projected examination date and work backwards. The following is a noncomprehensive example of a study plan based upon peaking out:

July 3--Projected examination date.
July 2--No study: if you don't know it by now, you'll never know it.
July 1--Review of difficult material. It seems that regardless of the study method, there is always some material that is difficult to remember, or that can be remembered only for a short period. This material should be identified and isolated during the study process and reviewed on this date.
June 24-26, 28-30 (June 27--no studying. One day a week should be reserved for relaxation.)--Review of previous examinations, Arco study manuals, etc. This review should reveal weak areas which should be covered during this period.

June 21-23--Rules and Regulations.
June 20--No studying
June 17-19--Supervision Principles.
June 14-16--Firefighting.
June 13--No studying.
June 10-12--Fire Code.
June 8-9--Hydraulics and Apparatus.
June 7--Arson
June 6--No study.
June 5--Hazardous Materials.
June 4--Miscellaneous material.
May 28-June 3--Break between study periods. No studying.
May 25-27--Rules and Regulations.
May 24-25--Supervision Principles.
May 23--No studying.
May 22--Supervision Principles.
May 19-21--Firefighting.
May 17-18--Fire Code.
May 16--No studying.
May 15--Fire Code.
May 13-14--Hydraulics and Apparatus.
May 12--Arson.
May 11--Hazardous Materials.
May 10--Miscellaneous material.

It should be noted that this is a systematic plan for the final study effort of all material gathered. It should not be considered the only time necessary for study. Much of the studying takes place before this final period. The schedule provides a cutoff date for gathering material and ensures that adequate time is allocated to learn the material. The sample plan is extremely comprehensive and is given for illustrative purposes only. Normally, a much longer period of time should be allocated for each subject area. The final study period should extend over approximately a three-month period.

Conclusion

Success on a written examination is determined by knowing what to study, then learning the material. It does not happen by chance. Success depends upon adequate preparation and planning, and much dedication and effort. There is no other way. The compensation will be rewards that are worth the effort.

SUPERVISION AND MANAGEMENT

Supervision

Questions 1 through 13 are based upon information contained in the text Managing Fire Services (John L. Bryan and Raymond C. Picard, editors, International City Management Association, Washington, D.C., 1979). This information has been used with the express permission of the International City Management Association.

1. The first and foremost objective of the fire department is

    (A) to serve, without prejudice or favoritism, all the community's citizens by safeguarding their lives collectively and individually, against the death-dealing and intrinsic effects of fires and explosions.

    (B) the safeguarding of the general economy and welfare of the community by preventing major conflagrations and the destruction by fire of large-payroll, economically essential industries and businesses.

    (C) to serve all of the community's citizens and property owners by protecting their individual material wealth and economic well-being against the destructive effects of fire and explosions.

    (D) to protect the citizens of the community against the destructive effects of war, particularly the rapid spread of war-caused fires.

2. Perhaps no specific management change offers the opportunity for increasing the productivity of paid firefighters more than

    (A) a change from the traditional working hours.
    (B) the use of firefighters for inspection work other than in the area of fire prevention.
    (C) increasing the time devoted to fire prevention and pre-fire planning
    (D) an increase in emphasis on training activities.

II

SUPERVISION AND MANAGEMENT

Supervision

Questions 1 through 10 are based upon information contained in the text Managing Fire Services (John L. Bryan and Raymond C. Picard, Editors, International City Management Association, Washington, D.C., 1979). This information has been used with the express permission of the International City Management Association.

1. The first and foremost objective of the fire defense community is

    (A) to serve, without prejudice or favoritism, all the community's citizens by safeguarding their lives, collectively and individually, against the death-dealing and injurious effects of fires and explosions.
    (B) the safeguarding of the general economy and welfare of the community by preventing major conflagrations and the destruction by fire of large-payroll, economically essential industries and businesses.
    (C) to serve all of the community's citizens and property owners by protecting their individual material wealth and economic well-being against the destructive effects of fire and explosions.
    (D) to protect the citizens of the community against the destructive effects of war, particularly the rapid spread of war-caused fires.

2. Perhaps no specific management change offers the opportunity for increasing the productivity of paid firefighters more than

    (A) a change from the traditional working tours.
    (B) the use of firefighters for inspection work other than in the area of fire prevention.
    (C) increasing the time devoted to fire prevention and pre-fire planning.
    (D) an increase in emphasis on training activities.

3. The budget process that goes from the bottom upward is classified as

    (A) a line-item budget.
    (B) a planning-programming-budget system (PPBS).
    (C) zero-based budgeting (ZBB).
    (D) a lump-sum budget.

4. Which, if any, of the following is not considered part of a public budgeting system?

    (A) The mechanical system, whereby accountants and others who are financially-oriented make sure that all the figures add up and are entered in the proper spaces on the correct forms.
    (B) The systemic technicians, who "massage" the mechanically derived data to assure that the "most efficient and effective" alternatives are chosen.
    (C) The political element, whereby the elected officials are "educated" by inputs from special interest groups, the general public, and the staff.
    (D) All of the above are considered as part of the public budgeting system.

5. It can be said that by far the most prevalent budgetary format of municipal government is the

    (A) lump-sum budget.
    (B) line-item budget.
    (C) performance budget.
    (D) program budget.

6. The budget process that requires a complete and orderly review of all governmental efforts, as well as an attempt to place them in priority order in view of current beliefs and needs is

    (A) program budgeting.
    (B) line-item budgeting.
    (C) zero-based budgeting.
    (D) lump-sum budgeting.

7. At the heart of the continuing controversy over recruitment of firefighting personnel lies test validation. This most nearly means that

    (A) the test examines for job-related skills, knowledge and abilities, and job behavior.
    (B) a firefighting candidate would receive nearly the same score on the examination process if taken a second time.
    (C) all prejudice has been removed from the examination process.
    (D) all culturally biased test items have been eliminated from the examination.

8. Performance appraisal in the fire service is a viable discussion
    item. First of all, it must be said that

    (A) performance must be centered on initial objectives.
    (B) the primary objective of any appraisal system is to
         provide feedback to management.
    (C) all appraisal efforts must be thoroughly documented.
    (D) all appraisal procedures must be developed around the
         MBO principle.

9. The single most important factor in producing and maintaining a
    high level of proficiency in any fire department is the

    (A) training program.
    (B) evaluation process.
    (C) management audit.
    (D) supervisor-employee relationship.

10. The most readily available and most commonly used means of report-
    ing emergencies to the fire department is the

    (A) commercial telephone system.
    (B) box alarm system.
    (C) still alarm.
    (D) box telephone system.

11. The more complex the organization, the more highly specialized
    the division of work, the greater the need for

    (A) strict line discipline.
    (B) inter- (or intra-) departmental council for cooperation.
    (C) finer division of supervision.
    (D) coordinating authority.

12. One method of organizing a complex operation is to subdivide the
    work into small units, with each worker specializing in one phase
    of the operation. Which of the following is a likely disadvantage
    of this subdivision of work?

    (A) The volume of work turned out will be reduced.
    (B) The quality of work will be lower.
    (C) The employees will require more training.
    (D) The work will be harder to coordinate.

13. In discussions of governmental personnel administration, the merit
    system is ordinarily contrasted with

    (A) career service system.
    (B) the Borstal system.
    (C) the spoils system.
    (D) the bipartisan plan.

14. "Definite mechanisms and routines must be provided to the end of causing all groups within an organization to function harmoniously toward the attainment of the common objective." According to this statement, a definite system must be established for the purpose of

    (A) specialization.
    (B) discipline.
    (C) coordination.
    (D) line and staff control.

15. Company officers become aware of the abilities of their subordinates and therefore know what to expect of them. However, one of the primary hazards of this expectation is that

    (A) the abilities of people change as they get older.
    (B) a person may change his habits and the change may be missed by the company officer.
    (C) people do not perform the same every day.
    (D) a person may become ill and it will go unnoticed.

16. A problem that confronts a new supervisor in relationship to his subordinates and which requires the exercise of an unusual degree of skill and diplomacy is

    (A) selection of competent personnel.
    (B) planning the work of each employee.
    (C) changing established ideas.
    (D) choosing personal friends.

17. The span of control, the number of persons who can be effectively supervised by an officer depends least on the

    (A) number of levels in the chain of command.
    (B) routine nature of the work.
    (C) ability of subordinates to work independently.
    (D) amount of time available for supervising each subordinate.

18. Of the following problems, the one that represents the greatest difficulty in the supervision of members of the Fire Department is

    (A) insufficient opportunity to gain a knowledge of the capabilities and deficiencies of subordinate staff members.
    (B) the absence of concrete and measurable indices of work performance.
    (C) the difficulty of identifying and maintaining satisfactory lines of authority in the agency.
    (D) channeling loyalties into the most desirable goals.

19. Although the methods and immediate aims of supervision should vary with the level and nature of each employer's function, the primary purpose is the same in most instances and can be said to be the

(A) improvement of employee performance.
(B) development of promotional material at all levels.
(C) attainment by each employee of his full capacity for self-direction.
(D) channeling of information through administrative lines.

20. The chief reason it is important for a supervisor to take prompt action upon requests and suggestions from subordinates is that

(A) the supervisor will be unable to organize his own work effectively if he accumulates a backlog of employee requests and suggestions.
(B) the likelihood of favorable action is greater when a decision is promptly made.
(C) such action maintains the good morale of his subordinates.
(D) the department will suffer economic loss from delayed adoption of suggestions.

21. The direct, immediate guidance and control of subordinates in the performance of their tasks is called

(A) planning.
(B) coordination.
(C) supervision.
(D) management.

22. All of the following are characteristic of a good fire officer except

(A) instructing the firefighters thoroughly.
(B) enforcing safety measures.
(C) directing, personally, each work detail.
(D) completing details at maximum practical speed.

23. One of the important leadership functions of a supervisor is to develop a feeling of belonging in those whom he or she supervises. In developing this feeling which of the following is the least important?

(A) encouraging participation by asking for advice and help on problems pertaining to work and production.
(B) providing plans and material for a steady workload.
(C) keeping the group informed on the progress of work.
(D) discovering the political, religious, and economic background of his subordinates so as to better understand them.

24. "A fire officer should be a leader whom his subordinates follow with enthusiasm." A competent officer should realize that this kind of leadership is most effectively based upon

(A) close observance of precisely formulated rules and regulations.
(B) diligent study by both firefighters and officers.

(C) respect and confidence of the firefighters.
(D) responsibility to the taxpayers.

25. If you are asked a technical question by one of your subordinates and you do not know the answer, you should

(A) tell the subordinate you do not know the answer but that you will get the information at a later time.
(B) attempt to answer it the best you can.
(C) tell the subordinate that you are too busy to explain at the moment.
(D) suggest that the subordinate not bother you with things that can be researched.

26. A firefighter whose most recent experience has been in salvage operations is transferred to the Engine Company which you command. The first thing that you should do is to

(A) explain the company organization.
(B) explain the detailed operation of your company.
(C) assign the firefighter to work immediately.
(D) find out how much he or she knows about your company's operation.

27. The main goal of any service rating program should be the

(A) recognition of good employee performance.
(B) recognition of poor employee performance.
(C) improvement of employee performance.
(D) objective evaluation of employee performance.

28. The existence of high morale in an organization is best evidenced when

(A) employees are willing to subordinate personal objectives to the organization's objectives.
(B) employee working conditions are favorable and salary scales relatively high.
(C) disciplinary cases are few and infrequent.
(D) the average length of service and tenure of employees is relatively long.

29. The best method of handling a firefighter who is chronically dissatisfied is to

(A) transfer the firefighter.
(B) reprimand the firefighter before others.
(C) suspension.
(D) discuss the problem with the firefighter in detail.

30. The wisest course of action for an officer to follow when first confronted with antagonism by a particular subordinate is to

(A) show annoyance to such a degree that the incident will not be repeated.
(B) compliment the subordinate on his or her good points.

    (C)   tell the subordinate that there is something wrong
          with him or her.
    (D)   make no comment and try to discover the underlying
          cause.

31.   Which of the following words most closely indicates the relationship of a staff officer in his dealing with line supervisors?

    (A)   authority
    (B)   advisory
    (C)   management
    (D)   supervision

32.   A company officer receives an order from his battalion chief which must be passed on to his company. The company commander does not like the order and knows it will be unpopular with the firefighters. He explains his feelings to the battalion chief who tells him to proceed as ordered. The company commander should

    (A)   explain to the firefighters that he does not like the
          order but they will have to follow it anyway.
    (B)   try to put the blame for the order on the chiefs at
          headquarters in order to protect his battalion chief.
    (C)   write up the order and place it on the bulletin board
          without comment.
    (D)   pass the order on as if it were his own.

33.   Assume you are a captain and that you find that you have unjustly reprimanded one of the firefighters in your company. The best procedure is to

    (A)   ignore the matter, but be more careful in the future.
    (B)   make no apology, but allow the firefighter's next
          offense to go unreprimanded.
    (C)   justify the reprimand on the basis of an earlier offense.
    (D)   readily admit your mistake to the firefighter.

34.   The principal argument in favor of filling top positions in any department by promotional examination rather than by open competitive examination is that this procedure

    (A)   heightens public interest in the examination.
    (B)   encourages "career service" within the department.
    (C)   assures that capable men will be recruited.
    (D)   assures the consideration of seniority in examining
          the candidates.

35.   Which kind of writ would a court be likely to issue in order to issue an order compelling a public official to perform some duty required by law?

    (A)   injunction
    (B)   habeas corpus
    (C)   quo warranto
    (D)   mandamus

36. In the past, fire department administrators have frequently urged excessive manpower and equipment, but the progressive fire officer knows that

    (A) the number of firefighters and the amount of equipment are not the most reliable index of the quantity and quality of fire service rendered.
    (B) less equipment is needed in departments that have a large force of firefighters.
    (C) drills and tactical practice eliminate the necessity for textbook study.
    (D) the more firefighters you have, the less need there is for fire department equipment to be entirely modern.

37. There is a continual need for reorganization in any department; however, the chief officers should realize that even though certain improvements should be made, reorganization generally should be accomplished

    (A) more or less gradually in an unfolding sort of way.
    (B) only after a plan is submitted to a vote of department employees.
    (C) before the beginning of the next fiscal year.
    (D) after the plan is unanimously approved by fire department personnel.

38. Good supervision is essentially a matter of

    (A) patience in training workers.
    (B) care in selecting workers.
    (C) skill in human relations.
    (D) fairness in maintaining discipline.

39. Which of the following functions should always be considered as being within the scope of supervisory responsibilities?

    (A) recruitment of well-qualified personnel
    (B) evaluation of employees' performance
    (C) classification of positions
    (D) reclassification of well-qualified employees

40. The chief reason a superior officer watches over the performance of the work under his or her direction is to enable him or her to

    (A) make accurate service ratings of subordinates.
    (B) submit complete and accurate reports of work progress.
    (C) help subordinates obtain the best possible results from their expenditure of time and effort.
    (D) select those subordinates eligible for promotion.

41. A realistic approach to effective supervision is to consider the importance of the individual, chiefly because

    (A) any group of workers, regardless of size, is composed of individuals.
    (B) individuals in a crew all have different backgrounds.

(C) injustices can be guarded against by using this approach.

(D) by appreciating individual differences a smoothly operating crew can be built.

42. In order to assure effective supervision in an organization, it is important for a supervisor beginning a job to know

(A) the history and progress of the organization.
(B) the progress of other agencies doing similar work.
(C) the nature and scope of his or her authority in each field in which he or she functions.
(D) the policy of promotion for the employees of the organization.

43. All but one of the following are considered sound principles to follow in leading subordinates. It is not considered a sound principle to

(A) be direct in actions.
(B) instill confidence and competition among subordinates.
(C) create a feeling of warmth between the supervisor and his or her subordinates.
(D) ignore petty grievances and allow them to work themselves out.

44. If a captain is able to admit failures, mistakes, or lack of knowledge, he or she usually

(A) has a sense of security in his or her own competence.
(B) lacks authority with his or her subordinates.
(C) is incompetent on his or her job.
(D) has been pushed ahead too fast.

45. If a fire officer uses his or her authority to obtain acceptance of his ideas, the cause most probably is that

(A) subordinates have little respect for their supervisor if he fails to use his authority.
(B) discipline depends upon subordinates' recognition of the supervisor's absolute authority.
(C) the introduction of new ideas is solely his or her responsibility.
(D) his or her ideas do not warrant consideration of their own merit.

46. If a situation arises in the line of duty which, to you, is not clearly covered by rules and regulations, you should

(A) use your best judgment in the matter.
(B) refer the matter to a superior officer.
(C) do as the majority of the firefighters think best.
(D) visit the battalion chief and consult with him.

47. "The problem of developing tests for the selection or evaluation of personnel is enormously complicated by the paucity of our information as to what we are trying to measure." In other words, it is difficult to find adequate

    (A) criteria.
    (B) samples.
    (C) test reliability.
    (D) classification techniques.

48. One of the most vital factors in developing employee morale within any relatively large organization is the

    (A) relationship existing between the employees and the head of the organization.
    (B) specific kind of production in which the organization is engaged.
    (C) policy of allowing employees great discretion in the quality of production.
    (D) relationship existing between the employees and their immediate superiors.

49. Which one of the following traits of a captain would be most likely to win and hold the respect of his company?

    (A) cheerful disposition
    (B) competence
    (C) decisiveness
    (D) courage

50. In the satisfactory handling of a complaint which is fancied rather than real, the complaint should be considered

    (A) unimportant since it has no basis in fact.
    (B) as important as a real grievance.
    (C) indicative of overpaternalism.
    (D) an attempt by the complainant to stir up trouble.

51. A company commander can best be described as a

    (A) supervisor.
    (B) manager.
    (C) planner.
    (D) all of the above.

52. A firefighter has made a mistake which the company commander believes deserves a reprimand. The primary objective of this reprimand should be to

    (A) punish the firefighter.
    (B) change the firefighter's behavior.
    (C) impress upon other firefighters the seriousness of not following orders.
    (D) make sure the offense is recorded.

53. There are certain tasks that are not liked. Because the work has to be done, and no other methods have been found, the supervisor should

    (A) assign the work as punishment for any failure to follow orders.
    (B) rotate firefighters on these assignments.
    (C) assign work to the healthiest firefighters.
    (D) assign work to the best firefighters.

54. That a subordinate should be under the direct control of one and only one immediate superior officer is a definition of the administrative principle of

    (A) control.
    (B) coordination.
    (C) span of control.
    (D) unity of command.

55. The basic unit of fire department organization is the

    (A) fire alarm system.
    (B) fire company.
    (C) records division.
    (D) fire prevention bureau.

56. In regard to reorganizational procedure in the fire department, the chief officers should constantly guard against

    (A) rotating personnel.
    (B) considering established organizational structure as being infallible just because of custom and long use.
    (C) any marked change from long-established practice that has been successful in the past.
    (D) making any substantial changes in organizational structure lest they be branded radicals.

57. Normally, operating personnel tend to resist and resent changes or innovations in existing procedures. To the supervising fire officer, the most desirable solution to this problem is to

    (A) hold the senior officer of each company responsible for the proper execution of the procedure.
    (B) plan the procedure so that it interferes least with current procedures.
    (C) develop acceptance by providing information concerning the procedure before its establishment.
    (D) describe the steps in the procedure in terms that will arouse least resentment among subordinates.

58. A generally accepted principle of supervision is that no individual can successfully handle more than a limited number of immediate supervisory contacts. This limit of control is principally a matter of the limits of

    (A) functionalization of duties.
    (B) work knowledge.
    (C) the kind of organization.
    (D) time and energy.

59. A full investigation of the facts should precede the solution of a supervisory problem primarily because

    (A) firefighters are usually critical of a battalion chief's decision.
    (B) an incomplete or unwise solution is likely to create new problems.
    (C) the evidence should be recorded for future reference.
    (D) the facts can be used to solve other problems.

60. Of the following, the <u>most</u> significant quality in relation to good fire department supervision is

    (A) the ability of the supervisor honestly to treat subordinates as whole human beings, with all possible combinations of human strengths and weaknesses.
    (B) the ability to give and take commands without question.
    (C) an understanding that the fire department is a semimilitary organization and that it must be conducted like the U.S. Army.
    (D) the ability of the supervisor to understand his own personal limitations.

61. "_____ consists of a study of a problem and the presentation of a solution in such form that all which remains to be done on the part of his immediate supervisor is to indicate his approval or disapproval of the recommended solution." The blank in the above quotation is best filled by

    (A) The scientific method.
    (B) Completed staff work.
    (C) Operations research.
    (D) Factor analysis.

62. Compliance with regulations should, whenever possible, be achieved through

    (A) an insistence on adherence to technicalities.
    (B) constant reminders concerning the consequences of non-compliance.
    (C) teaching subordinates to like and respect you.
    (D) a demonstration of the purpose and value of the regulations.

63. A subordinate officer presents you with a long-standing problem which he has analyzed in detail. However, he states that he has no idea of its solution, nor can he make any recommendations. Under these circumstances, the most appropriate action for you to take at this time is to

    (A) analyze the problem with the officer and suggest a number of alternative solutions from which the officer is to pick the most suitable.
    (B) tell the subordinate officer that you will not tolerate "buck passing."
    (C) refer the problem for conference review to determine the most effective approach to its solution.
    (D) suggest that the subordinate officer review the matter and recommend possible actions which might solve the problem.

64. If a firefighter under your supervision disagrees with your evaluation of his work, the most desirable way for you to handle the situation would be to

    (A) suggest that he take the case to a higher authority
    (B) explain that you are in a better position than he to know whether or not his work is up to standard.
    (C) suspend him for ten days.
    (D) discuss specific details with him, showing where improvement is necessary.

65. When a fire captain takes time to so instruct subordinates that they understand the why and wherefore of their duties, he or she is

    (A) making friends.
    (B) developing morale.
    (C) wasting time.
    (D) being dictatorial.

66. Which of the following should be the supervisor's attitude toward grievances?

    (A) Know the most frequent causes of grievances and strive to prevent them from arising.
    (B) Pay little attention to little grievances.
    (C) Maintain rigid discipline of a nature that "smothers out" all grievances.
    (D) Be very alert to grievances and make adjustments in existing conditions to appease all of them.

67. Of the following, the best general rule for a superior officer to follow is to assign work, wherever possible, on the basis of

    (A) individual abilities.
    (B) expected time to perform the job required.
    (C) previous assignment.
    (D) seniority.

68. An oral reprimand to a company officer should be given

    (A) immediately at the time the situation arises.
    (B) in the company of his subordinates in order for it to be effective.
    (C) in private.
    (D) never, as an officer should receive only official reprimands.

69. Suppose that as a company commander you have assigned a task which will take several hours to a firefighter under your command and have given him appropriate instructions. About a half-hour later, you check the progress of his work and find that one specific aspect of his work is consistently incorrect. Of the following, the best action for you to take under these circumstances is to

    (A) repeat your instructions to the firefighter in full and then check the progress of his work again about a half-hour later.
    (B) assign that task to a firefighter who you believe will perform the work more competently.
    (C) determine whether the firefighter has correctly understood your instructions concerning the specific aspect of the work not being performed correctly.
    (D) observe the firefighter at his work carefully for a brief period of time to determine whether you can detect the reason for his mistakes.

70. When it is said that the functions of management may be divided between planning and control, the term "control" means

    (A) restricting the powers of those who plan the policies of an organization by independent review of all policies.
    (B) making sure that the policies of the organization are carried out.
    (C) determining the specific objectives of the organization.
    (D) establishing responsibility for planning.

71. One of the principal disadvantages of a strictly line organization is that it

    (A) is ineffective in emergencies.
    (B) makes for lax discipline.
    (C) does not fix responsibility.
    (D) tends to overload executives.

72. Which one of the following is the primary objective in drawing up a set of specifications for materials to be purchased?

    (A) control of quality
    (B) establishment of standard sizes
    (C) location and method of inspection
    (D) outline of intended use

73. "Probably the most important part of supervision is to stimulate subordinates to want to do the required work and to prepare themselves to perform more and better work." Which one of the following is the most important part of this kind of supervision?

    (A) rewarding superior work
    (B) carefully planning all work
    (C) satisfying the needs of subordinates
    (D) conducting training courses

74. Assume that a new evolution has been set up for a new piece of equipment. The first few times that the evolution is used at a fire, a close check should be kept on all operations. The primary reason for this is to

    (A) learn whether the officers and men have accepted the evolution.
    (B) determine the efficiency of the officers and men.
    (C) emphasize the special values in using this equipment.
    (D) determine if the evolution requires modification.

75. Assume that you have been appointed a captain. Which of the following is the best justification for learning from the battalion chief as much as possible about the firefighters whom you are to supervise?

    (A) Personality problems usually disappear with knowledge of individual differences.
    (B) Knowledge of the individual characteristics of the firefighters often aids in the effective handling of them.
    (C) No captain can be effective in his work unless he has a cooperative relationship with his or her superior.
    (D) Such knowledge will provide the basis for mutual understanding between the captain and his firefighters.

Answers

1. (A) <u>Managing Fire Services</u> gives three objectives for the fire defense program of a community. The first and foremost objective is given in answer (A). The second most important objective is given in answer (B). The third objective is given in answer (C).

2. (A) Perhaps no specific management change offers the opportunity for increasing the productivity of paid firefighters more than a change from the traditional working tours, which include sleeping time, standby time, etc. The twenty-four-hour tour of duty builds in substantial amounts of nonproductive nonemergency time, even if eight hours of productive nonemergency are programmed into every on-duty shift. The split ten-hour day fourteen-hour night is an improvement over the twenty-four hour shift in that perhaps sixteen hours each can be devoted to productive nonemergency work: the day shift works and then goes off duty, then the night shift comes on, works, and goes to bed.

3. (C) The PPBS is a process of budgeting from the top downward. Zero-based budgeting, on the other hand, is a budgeting process that goes from the bottom upward. It establishes goals and objectives and measures results as part of the budgeting cycle. In ZBB, each year at budget time an organization is required to justify its ongoing as well as its new and proposed programs and its funding levels from scratch. ZZB is intended to weed out programs which are unnecessary or are no longer useful. It is a painful task for most managers to have to justify the empire so carefully built, but the exigencies of public funding demand that this be done.

4. (D) A public budgeting "system" consists of the three choices given. The three are disconnected but carefully intertwined elements. Such is the setting of any--and every--budget system, no matter how small, or how gigantic, the jurisdiction.

5. (B) By far the most prevalent budgetary format, and the original "reform" budgetary format, the line-item budget is basically a listing of items of purchase and acquisition, and personnel employed by department, by agency, or, in some cases, by a lesser organizational unit within the larger component. The concentration of objects of expenditure (line item) in considerable detail often results in a lengthy document.

6. (C) This is one of the so-called advantages of zero-based budgeting. Furthermore, exploration of new ways to accomplish old objectives is encouraged, as is an inherent decentralization of authority and decision making.

7. (A) At the heart of the continuing controversy lies test validation. Test validation means that documented proof exists that the test does in fact examine for job-related skills, knowledge and abilities, and for job behavior. Validity may be stated as: Does the test in reality measure what it was intended to measure? One does not measure weight with a yardstick. Associated with the concept of validity are the equally important subjects of job task analysis and cultural impartiality.

8. (C) First, all appraisal efforts must be thoroughly documented. Second, both supervisory and line personnel should read the guidelines once a year. Third, the raters should be periodically trained and evaluated in appraising an employee's performance. Fourth, multiple raters should be used to evaluate each person. Fifth, the rating form should be job-related. Sixth, the employee should be required to set individual objectives for accomplishment during the subsequent time period.

9. (A) In <u>Managing Fire Services</u>, Anthony Granit makes the following statement:

    "A good training program is undoubtedly the single most important factor in producing and maintaining a high level of proficiency in any fire department. It not only produces high efficiency initially, but also affects future efficiency when we consider that the rawest recruit now being trained may be chief of department or at least a senior officer in 20 or 30 years."

10. (A) The commercial telephone system is the most readily available and commonly used means of reporting emergencies to the fire department. Regardless of its availability and widespread use, however, the telephone system should be considered a secondary rather than primary means of fire alarm service because of its vulnerability to disruption of service. An earthquake, tornado, or other natural disaster may disable telephone service, as may such "man-made" interruptions as accidents, sabotage, or even a strike of telephone company employees.

11. (D) Coordination might be defined as arranging work so that it fits in proper perspective with the work of others. In large organizations, and in organizations where the work is divided into highly specialized areas, there is a tendency for those responsible to become so engrossed in their own particular operation that they lose sight of overall organizational objectives. When this happens, there is a strong need for one individual to have the authority to coordinate all activities toward the common goal.

12. (D) There are certain advantages in a complex operation to making each worker responsible for a specialized portion of the job; however, this compartmentalization tends to create a situation where each worker loses sight of the overall objective of the operation. This makes it much more difficult to coordinate all the activities toward the common goal.

13. (C) The principle of the merit system is to hire the best qualified person for the job by making every candidate for a position compete in the examination process. In theory, it creates a government run by professionals.

    The theory of the spoils system, on the other hand, is based upon the concept "to the victor goes the spoils." This means that those who win elections can appoint people to government positions regardless of their qualifications. This could create a government run by amateurs.

14. (C) The work in an organization is normally divided up among workers or units according to some logical plan. Some of the work may be classified as line functions and some as staff functions; however, a definite mechanism must be established to ensure that all work is coordinated in order to direct it toward a common goal.

15. (B) All of us have a tendency to anticipate how people we know will perform in a particular situation. Although this expectation has certain advantages in supervision, it also can be dangerous. Many times a supervisor is so certain how one of his subordinates will perform in a given situation that he completely overlooks the fact that performance differed from that anticipated.

16. (C) Supervisors should recognize that making changes requires an unusual degree of skill and diplomacy; however, changes can be made successfully. In making changes, supervisors should keep in mind that: 1) people resist change, 2) changes should be made gradually, and 3) people should be informed of the need and reason for a change before its initiation.

17. (A) Span of control refers to the number of persons that can effectively be supervised by an officer. Span of control depends upon a number of factors such as the nature of the work, amount of planning and control required, the abilities of subordinates, amount of time available for supervision, the proximity of the people being supervised, etc. However, the span of control is not affected by the number of levels in the chain of command.

18. (B) It is easy to judge the efficiency of a worker in industry when a certain number of units are required to be produced each hour and definite criteria are available to judge the quality of the units. The fire service is different. The determination of how well a firefighter does the job is usually based upon subjective judgment. As an example, what are the criteria for determining how well a line is laid? What are the criteria for determining how well a ladder is raised? How clean is clean? This lack of concrete and measurable indices of work performance in the fire service makes it extremely difficult to judge the efficiency of personnel and to establish criteria for training performance.

19. (A) Supervisors utilize different methods and techniques to get the work done; however, their ultimate success depends on how well the subordinates in their command perform. A supervisor is only as successful as the subordinates make him or her. Regardless of his management style, it should be the primary responsibility of every supervisor to do whatever is necessary to improve the performance of the people who work for him or her.

20. (C) Morale may be defined as a spirit of devotion to the objectives of a group. One of the primary factors affecting the morale of an individual is the degree of the feeling of belonging to the group. This feeling of belonging carries with it the importance of recognition and acceptance. Any action taken by a supervisor which makes a worker feel that his or her ideas or suggestions are not important results in the worker's losing the feeling of belonging. It is therefore important for supervisors to take prompt

action upon requests and suggestions from subordinates, even if he or she believes that a request or suggestion is of little importance.

21. (C) Some people say there is little difference between managing and supervising; however, the gap is wide. A person can manage an organization without coming in direct contact with people, but supervision, by its very nature, cannot exist without people. Supervision exists when there is direct contact with people in a face-to-face relationship, with the supervisor guiding and controlling the performance of the individuals.

22. (C) A supervisor who gives a worker a job to do and then permits the worker to figure out the best way to do it instills in the worker a sense of worth. Of course, the worker must be provided the training required to get the job done, and the supervisor must be available to help, if help is needed, but the worker must be given the leeway to accomplish the job without the constant direct supervision of his or her superior. A supervisor who tries to direct and personally supervise every work detail becomes a snoopervisor rather than a supervisor.

23. (D) All the selections except (D) will assist in the development of a feeling of belonging. Although a supervisor should know his or her subordinates and their individual differences in order to understand their behavior and needs, delving into their political, religious, and economic backgrounds can be considered an invasion of privacy. In addition, these factors should have no effect upon the manner in which a subordinate is supervised.

24. (C) Respect is earned; it does not go automatically with a rank. Although leadership is difficult to define, it certainly indicates that the person holding the title has the ability to lead. Subordinates hesitate to follow anyone in whom they have little confidence. Nowhere is this more true than in the fire service, where the lives of those following orders depend upon the knowledge and skill of those in command. It almost goes without saying that a company officer must have the respect and confidence of his or her subordinates.

25. (A) It is extremely important for subordinates to have respect and confidence in their company officers. Confidence is built upon a foundation of dependability. Subordinates must be able to depend upon the information given to them by their officers. Whenever an officer is asked a question to which he does not know the answer, he or she should inform the subordinate of the fact and tell the subordinate that he or she will obtain the information. Moreover, it is very important that the officer follow through on this.

26. (D) The training need of an individual is the difference between what he knows and what he needs to know to get the job done. For lack of a better term, this difference might be referred to as a knowledge or skills gap. The responsibility of a company officer is to eliminate this gap. Although he knows what a newly assigned firefighter must know to get the job done, he cannot determine the

width of the gap until he discovers what the firefighter knows in
relation to what he must know.

27. (C) The main goal of any service rating program should be the improvement of employee performance. Unfortunately, this is not the
way it works in actual practice. Both workers and supervisors dis-
like service ratings. Workers hate to be rated, and supervisors
hate to rate workers. Consequently, the most common errors evident
in most rating systems are leniency and the tendency to rate every-
one from average to above average.

28. (A) When the employees within an organization willingly and en-
thusiastically work toward a common goal the morale is said to be
high. Because the needs and desires of individuals within an organization are generally different than those of the organization,
it is necessary for individuals to subordinate their personal needs
and desires to those of the organization. Employees will not willingly do this when morale is low.

29. (D) Get the facts. Find out the reason for the dissatisfaction.
When doing this, it is important to realize that it may be necessary to talk to the firefighter for some time before the facts behind the dissatisfaction surface. The first reasons given are
usually not the real reasons, and the company commander can do
little to change the situation unless the facts are known.

30. (D) Let the subordinate talk. Many times this in itself will
solve the problem. It might be a matter of the subordinate's having a need to get something off his or her chest. Even if this is
not the answer, the problem cannot be solved until the subordinate
calms down. Once this is accomplished, then the officer should try
to discover what is causing the unhappiness. To solve any problem,
it is first necessary to get the facts.

31. (B) A staff officer has no authority in dealing with line department heads. The officer can advise, counsel, suggest, and recommend, but he or she cannot give any direct orders.

32. (D) Not all orders that come down from above are popular; however,
it is the responsibility of the company officer to pass the order
on as if it were his or her own. One should not pass the buck or
blame on to higher authorities. Of course, if the reason for the
order is known, it might help in selling it to the subordinate.

33. (D) Anyone can make a mistake, and most people do. A mistake
which is not corrected, however, becomes a bigger mistake. If a
company officer unjustly reprimands a member of his or her company,
he or she should have a private talk with the individual and admit
the mistake. He or she should also compensate for the mistake, if
such action is necessary. Many times the action of admitting the
mistake is all the compensation that is required.

34. (B) Most people working within an organization have hopes of
eventually utilizing their skills and knowledge in higher positions
within the organization. They remain with an organization because
of the opportunity of moving up the ladder. Opening up promotional

positions to the outside normally results in lowered morale and, in time, a reduced efficiency within the department.

35. (D) Injunction is a writ forbidding a person from performing a non-criminal act.

    Habeas corpus (produce the body) is a writ to show cause why a person should not be released.

    Quo warrant (by what authority) is a proceeding where the right of a person to hold office can be challenged.

    Mandamus is a remedial writ. It is used to compel action from a public official.

36. (A) The efficiency of a fire department depends upon its ability to extinguish a fire with the least possible loss of life and property. Size alone is not a mark of efficiency. A well-trained team of firefighters with adequate equipment can usually produce better results than a poorly trained, well-equipped team with more personnel.

37. (A) Every supervisor should be aware that people resist change. Yet, there is nothing constant but change. The question itself implies this by stating, "there is a continuous need for reorganization in any department." Consequently, those responsible for making changes should lessen the impact of change by making changes gradually over a period of time.

38. (C) The skill of any worker depends on how well he or she is able to use the tools required to get the job done. The tools of a supervisor are people. Although all people are basically the same, no two are exactly alike. Good supervision is essentially a matter of working successfully with individual differences (human relations).

39. (B) The recruitment, classification, and reclassification of personnel are functions of the Personnel (or Civil Service) Department; however, the evaluation of subordinates is a direct responsibility of a supervisor.

40. (C) Every company officer is responsible for determining the training needs of his or her company. The training need might be defined as the difference between what subordinates need to know to do the job effectively and what they already know. A company officer should evaluate the performance of the personnel in his or her command in order to close this training gap by helping subordinates obtain the best possible results from their efforts.

41. (D) Knowing the people assigned is the first requirement of every supervisor. Workers differ mentally, physically, and emotionally. Even the same worker's abilities and interests varies from day to day. It is important for a supervisor to accept and appreciate the individual differences of personnel. Only through this acceptance and appreciation can the supervisor weld a group of individuals into a smooth, operating team.

42. (C) One of the first acts of any officer assigned to a new position is to gain a thorough understanding of the laws and rules and

regulations under which he or she will operate, his or her responsibilities, and the limits of his or her authority. This knowledge will provide the supervisor with the perimeter within which he or she must operate, and the foundation for establishment of an effective organization.

43. (D) A grievance may be defined as anything in the work environment that a worker feels or believes to be wrong, unfair, or unjust. An untreated grievance will fester and grow, eventually becoming a big sore. A supervisor must remember that a grievance is real, whether or not it has a factual basis. Supervisors must constantly be on the alert for grievances, treating them immediately, regardless of how petty they may seem.

44. (A) People who cannot say "I don't know," or "I made a mistake" are usually insecure. This insecurity in an individual is usually based upon fear--fear of reprisal for not knowing or making a mistake, or fear that others will consider that he or she is not qualified for the position held. A truly educated and secure individual is one who realizes how little he or she knows.

45. (D) Firefighters have little difficulty accepting ideas or following orders they understand and which appear sound. They hesitate, however, when the requested behavior seems unwise. Whenever an officer has to use authority to get subordinates to accept an idea, it is probably because the idea seems unsound and will not receive acceptance on its own merit.

46. (A) There is an important guideline that is found in the preface of many sets of Rules and Regulations. It goes something like this: These Rules and Regulations were not conceived nor intended to cover every situation that might arise in an emergency service. Much has to be left to the intelligence and dedication of the officers concerned. In plain language, this says, "When all else fails, use common sense."

47. (A) Webster defines criterion as "a standard of judging; any established law, rule, principle, or fact by which a correct judgment may be formed." The key in this definition is "by which a correct judgment may be formed." As an example, what standard can be established to judge whether or not an employee is honest or loyal? What standard can be used to determine if a firefighter performs well at emergencies? What standard can be used to determine if an employee's performance is below average, average, or above average in his daily work? Where should the lines be drawn between good, adequate, satisfactory and excellent behavior?

48. (D) There is nothing that has more of an effect upon the morale of a fire company than the relationship between the firefighters and the company officer. If the firefighters respect and have confidence in their officer, and the officer treats subordinates fairly, equitably, and as individuals, the morale of the company will probably be high. This is true many times even when outside influences have an adverse effect. Although the subordinates may revolt against the outside influence, their morale as a company is generally high.

49. (B) Although this is true in most organizations, it is probably even more so in the fire service. Of course, the word competence does not apply to firefighting operations only. It also applies to the manner in which an officer carries out responsibilities in public relations, fire prevention, and other aspects of fire protection.

50. (B) A grievance is a sign that something is bothering a person. Whether the cause of the grievance is factual or imaginary, the results are the same. Grievances are problems requiring immediate action, regardless of how trivial the complaint may be. Little irritations which manifest themselves as grievances develop into large problems if not handled promptly and efficiently.

51. (D) A company commander wears many hats. He or she is a supervisor, manager, and planner, all rolled into one. As a supervisor, he or she directs the firefighters in his or her company. As a manager, he or she conducts the company business. As a planner, he or she plans company operations.

52. (B) Having to reprimand a firefighter tests the caliber of the company commander. It also provides an opportunity to improve a worker's performance. The objective should not be punishment, but rather to change the firefighter's behavior and get him or her to think in terms of insuring that the action will not take place again. It is an occasion calling for training, not blame.

53. (B) There are always tasks that have to be performed that are not particularly popular. It is important for a company commander to rotate these tasks among members of the company, making sure favoritism is not shown to any member, and that no member gets more than his or her share of the dirty jobs.

54. (D) Unity of command is the concept that a person can work successfully for only one boss at any given time. Those receiving orders from several superiors become confused and inefficient.

55. (B) The basic unit of a fire department is the fire company. A company normally consists of a company officer, an apparatus operator, and one or more firefighters. The engine company is the most common kind in the fire service.

56. (B) All officers must guard against complacency. Fire departments are steeped in tradition; however, whenever reorganization is being considered, tradition must be set aside and procedures judged on their merits. Once procedures have been evaluated against well-defined criteria, then tradition can be considered.

57. (C) Question 37 discussed change, stating that it is normal for people to resist change. Although it is normally better to initiate change gradually, acceptance can best be achieved by proper communication. People resist change, but they accept it more willingly when they are aware of the need and recognize that results will probably be better. It is therefore much better to provide as much information as possible regarding a change before its implementation.

58. (D) Span of control refers to the number of people that can successfully be supervised by one person. The span of control of a supervisor varies depending upon factors such as the kind of work being done the physical location of the people being supervised, the amount of training required, etc. However, in essence, the span of control of a supervisor depends upon the time and energy he or she has to get the job done.

59. (B) There is a saying that nothing can be sliced so thin that there are not two sides to it. It is easy to arrive at the wrong conclusion if both sides are not seen. This is true of any problem. Not having all the facts could result in arriving at the wrong conclusion; rather than solving the problem, a bigger one could result.

60. (A) A sign on a building at the University of Southern California states in essence that education is learning to use the tools that man has found to be indispensable. The tools of a company commander are the subordinates. He or she has to learn to work with these tools, understanding that each is different, having different strengths and weaknesses. Only by compensating for these individual differences in the treatment of his or her people can a company commander be successful in supervising a company.

61. (B) Functions in an organization may be divided into line and staff. Those performing line functions carry out the basic work of the organization, while those performing staff functions provide service to the line organization. Completed staff work consists of the study of a problem and the presentation of a solution in such form that all that remains to be done on the part of the supervisor is to stamp his approval or disapproval on the recommendation. No further study nor research has to be made.

62. (D) Most people do not object to complying with rules and regulations if they understand the reasons for them and why they are needed to carry out the basic objectives of the organization. Lack of understanding of a regulation, however, will cause workers to resist complying, if for no other reason than it is inconvenient to comply with a rule that is unnecessary or unsound.

63. (D) It is not uncommon for a subordinate officer to try to get a superior to make a decision which he or she is reluctant to make. This is particularly true when it comes to disciplining employees. It is unwise, however, for a superior officer to make a decision that should be made by a subordinate.

Normally, in this sort of situation, the junior officer will approach the senior for assistance, stating he has a problem and would like a recommendation on how it could be handled. The wise superior will ask the subordinate what he or she thinks are possible solutions, and when the subordinate suggests several, tell the subordinate to take the alternative which seems to be best.

64. (D) It should be kept in mind that the primary purpose of a rating system is to improve employee performance. If an employee being rated has been falling down in work performance, a general note to

this effect is not sufficient. To improve performance, an employee must be given specific examples of unsatisfactory work performance, including date, time and place, and must thoroughly understand the standards by which he or she is being measured, and must be shown different ways that his or her performance may be improved.

65. (B) Of all the factors affecting morale, the relationship between an officer and subordinates is the most important. The value of an officer taking time to instruct subordinates so that they understand their duties thoroughly will be appreciated. The supervisor is indicating an interest in their general welfare and is treating them as individuals, making them feel they are part of the team. They are being stimulated in a manner that will get them to work willingly toward the common company goals.

66. (A) Grievances are much like fires. Although it is important to have the knowledge and skills necessary to extinguish them, it is much better if we devote time and effort to preventing them from starting in the first place. To prevent fires, we must know their causes and then establish and implement a plan to prevent them from starting. So it should be with grievances. Knowing their causes and striving to prevent them is the proper approach.

67. (A) An important part of the company commander's responsibility is matching the ability of available personnel to the job to be done. To do this, the company commander must be thoroughly familiar with all required work, and must know the strengths, abilities, and limitations of each member of the crew.

68. (C) Oral reprimands should be given in private. If given at the time of the occurrence of the infraction or in the company of other members, it will do nothing but embarrass the person receiving it. It should be remembered that reprimands should be used as training situations with the objective of changing the behavior of the individual in such a manner that the action requiring the reprimand is not likely to occur again.

69. (C) Remember, workers do not purposely perform tasks incorrectly. If a worker has been trained to do a job, and is performing it incorrectly, most likely, the training was not unterstood. A good method of evaluating whether or not workers understand instructions is to have them repeat the directions given in their own words.

70. (B) As a management term, control means the establishment of methods and procedures, to ensure that the objectives of an organization are being achieved. In the establishment of methods of control, it is essential to understand the objectives and to develop methods of securing information regarding the progress being made toward their achievement. Two of the primary tools of control are reports and inspections.

71. (D) A strictly line organization is a simple organization and one which is easily understood; however, it tends to overload some of the key executives as it does not provide for specialists who can be used as advisors. The absence of these specialists requires

executives to spend time on research which should be used to achieve the basic functions of the organization.

72. (A) Drawing up a set of specifications in essence establishes criteria. When goods are received, they can be measured against these criteria which provide for a method of quality control. Without specifications (criteria), there would be no way of measuring the quality of material received.

73. (A) People work much better when they receive recognition for work accomplished. Although a pat-on-the-back for doing a good job on a day-to-day basis has merit, recognizing and rewarding personnel for superior work has lasting benefits.

74. (D) A new method or evolution is not set in concrete when first implemented. Even though much thought goes into the development of a new method, there is no assurance it will work as well as planned. New operations must be watched and evaluated to determine if the procedure is satisfactory as implemented, or if modifications are needed for improvement.

75. (B) A worker assigned to an apparatus would be foolish to attempt to operate the apparatus until he or she thoroughly understood its operation. The knowledge required provides the foundation for effective performance. The tools of a supervisor are the personnel. He or she must also have knowledge regarding the individual differences of his or her people in order to build the team desired. Knowing one's subordinates provides the basis for understanding their actions, and will also provide the basis for subordinates to understand their supervisor.

Training

1. Studies of the learning process indicate that the speed of absorption of new skills is not constant. Plateaus, or periods in which little or no progress is made, are regularly encountered. The most important implication of these findings for the officer conducting a training program is that

    (A) training sessions should be spaced to avoid the plateau period.
    (B) trainees should be encouraged during the plateau period.
    (C) before the training program starts, trainees should be warned to expect the plateau period.
    (D) training efforts should be concentrated on the plateau period.

2. Every company commander who has had occasion to teach his men how to operate a new piece of equipment has seen trial-and-error learning, in which the firefighter fumbles about until he or she strikes upon the proper procedure by accident. Of the following, the most accurate statement concerning trial-and-error learning in fire training is that

    (A) trial-and-error should be reduced by the company officer through proper guidance.
    (B) the company officer will find it most effective to allow trial-and-error learning to precede specific training.
    (C) trial-and-error learning is more permanent than any other kind of learning.
    (D) trial-and-error learning is more efficient per unit time than any other kind of learning.

3. Assume you have been assigned to train a group of officers on some new apparatus. Which one of the following criteria should be given primary consideration when planning your course of instruction?

    (A) How much time will I have at my disposal?
    (B) How much will this cost?
    (C) What should the officers learn in this course?
    (D) What is the background of the officers?

4. Of the following, which is the most important reason for evaluating the outcome of a training program by an officer?

    (A) More active participation may be obtained from company members when such evaluations are shared.
    (B) The introduction of new or improved firefighting techniques is facilitated.
    (C) Shortcomings discovered can serve as the basis for further training.
    (D) Success of the program is evidence of qualities of leadership.

5. The expenditures needed to maintain a training program for a large fire department are fully justified only if the

    (A)   firefighters' morale is improved.
    (B)   instructors acquire knowledge over and above that required to do their routine work.
    (C)   employees' knowledge is noticeably increased.
    (D)   department's efficiency is noticeably improved or maintained at a high level.

6.   In leading a training conference, it is usually best to make your introductory remarks, give the first question for discussion, and

    (A)   call on a group member for response.
    (B)   ask for contributions of ideas from members in rotation around the table.
    (C)   let the group members respond voluntarily.
    (D)   discuss the question and its problem at some length yourself.

7.   A new driver develops a tendency to overspeed the engine before shifting gears. As the driver's superior, the most effective measure for you to take would be to

    (A)   report the member's deficiency to the Division of Safety.
    (B)   request that the driver be relieved of his assignment.
    (C)   institute corrective training emphasizing both theory and performance.
    (D)   discuss this condition with the driver to get at its cause.

8.   A company commander is teaching one of the members of his company how to use a new piece of equipment. After explaining the proper procedure, he has the individual try it. He then has the individual repeat the performance and explain each step. In doing so, the person makes a mistake. The company officer should

    (A)   say nothing.
    (B)   wait until the individual has completed the entire process then point out the mistake.
    (C)   stop the individual immediately before the wrong method becomes set.
    (D)   have another member of the company later tell the individual where he or she had made a mistake.

9.   When presenting a completely new idea to a group of workers, the most important principle of learning to be observed is to

    (A)   relate the new idea to some familiar idea or activity previously learned.
    (B)   point out how difficult the new idea is to learn.
    (C)   give the employee printed material to read on the new idea.
    (D)   break down the new principle into constituent parts before presenting the whole.

10.   Learning is a matter of using as many of the senses as possible. Of all the senses, the one contributing the most to learning is

(A) seeing.
(B) hearing.
(C) smelling.
(D) feeling.

11. A company is faced with a troublesome local problem. The company commander calls all the lieutenants to a meeting, outlines the problem and asks them to give spontaneously any ideas that occur to them as possible ways of handling it. Each idea suggested is written down, and later discussed carefully. The chief advantage of the procedure employed by the company commander is that

(A) time is not wasted on needless talk.
(B) ideas are obtained which otherwise might not be developed.
(C) there is less tendency for the meeting to stray from the subject under discussion.
(D) lieutenants receive training in analysis of problems and evaluation of solutions.

12. Following are the steps in the four-step method of instructing. If given in proper sequence, the third step would be

(A) present.
(B) follow-up.
(C) perform.
(D) prepare.

13. The staff conference, as a training method, can be used <u>least</u> effectively when

(A) the backgrounds of the participants are diversified.
(B) there is a wide difference of opinion concerning questions to be discussed.
(C) the participants are experienced in the problems to be discussed.
(D) the subjects discussed pose no problems.

14. Of the following, the training method likely to be most effective in developing a specific skill on the part of firefighters is

(A) well-planned lectures.
(B) selected readings.
(C) interesting demonstrations.
(D) repeated practice.

15. It is the responsibility of every instructor to correct the mistakes made by trainees. However, of the various methods of correcting errors the best is to

(A) compliment before correcting.
(B) strongly tell the trainee that a mistake has been made.
(C) show the trainee how he could have done it better.
(D) let the trainee correct himself.

16. "Any individual thoroughly familiar with the work of a firefighter is well-qualified to serve as an instructor in a fire academy." This statement is not correct chiefly because

(A) no one man can know everything there is to know about a firefighter's work.
(B) a qualified instructor should not be expected to be familiar with the work of a firefighter.
(C) the skills and knowledge that a firefighter must have are not independent but are interrelated.
(D) the quantity of information possessed by an instructor does not bear a direct relationship to the quality of instruction.

17. When a company officer finds that many of the firefighters under his or her command make numerous mistakes on work which they should be able to do, the most probable cause for the mistakes is that the subordinates

(A) have poor aptitude for the work.
(B) have had insufficient training for the job.
(C) are doing work that is too difficult.
(D) have duties that are too varied.

18. A conscientious supervisor, in conducting on-the-job training, will normally find which of the following to be the greatest challenge to his supervisory abilities?

(A) preparing material for presentation
(B) instilling a desire to learn in the employee
(C) presenting the material to be learned
(D) having the employee demonstrate what has been learned

19. Which one of the following is the most essential factor in a training program for a large fire department?

(A) adequate drill tower facilities
(B) adequate supply of written material on firefighting
(C) a knowledge of what other large fire departments are doing in their training programs
(D) officers who know firefighting and who have a good knowledge of instructional methods

20. The best measure of the results of a formal training program is

(A) final test scores achieved by the trainees.
(B) differences in performance of duty between personnel who took the training course and those who did not.
(C) the grades of trained personnel in their next promotional test.
(D) differences in performance of duty of the personnel trained, before and after training.

Answers

1. (B) Most students seem to progress well in their training efforts and then hit a learning plateau where little if any progress takes place. The length of time students remain on this plateau varies; however, there is a tendency during this period for students to become extremely discouraged. Many students quit during this period. It is the responsibility of an instructor to recognize that a student is on a plateau and to do his or her best to encourage the student through this difficult period.

2. (A) Trial-and-error learning is a poor teaching method. Not only is it time-consuming, but students can develop bad habits using this method. Additionally, without constant supervision, trial-and-error learning can be conducted in an unsafe manner. Students should be allowed to make mistakes by which they grow, but the mistakes should normally only be allowed after the student has been instructed and guided in the proper method of performing the operation.

3. (C) Whether an instructor is developing an entire training program or merely a training lesson, the first step should be to establish the objectives--in other words, what skills or knowledge should be acquired from the training. Although this principle of training is logical and simple, it is surprising how many times company training sessions are conducted without the company officer establishing an objective for the training.

4. (C) The first step in the development of a training program is to establish the objectives (what is to be learned). After the program has been conducted, it is then necessary to measure the knowledge and skills of the learners to determine if the objectives have been met. If shortcomings are discovered during the evaluation process, this information should then be used as the basis for subsequent programs or training sessions.

5. (D) Whatever else it might be called, training is change. All training should result in a change of habits, change of knowledge or attitude, or change of skills of participants. If no change occurs, no training has taken place. Consequently, it is difficult to justify expenditures for training programs if employees do not exhibit a profitable change (in performance) upon completion of the training.

6. (C) The responsibility of the person leading a conference training session is to encourage participation and to secure the contributions of those at the conference. In the opening of the conference, the conference leader should state the purpose and objectives of the conference, define the procedures to be followed in the discussion, and then state the initial problem to get the discussion started.

7. (C) A driver who has a tendency to overspeed the engine before upshifting gears is probably not aware of exactly what happens within the gear box when shifting of gears takes place. The driver must be taught what goes on inside the transmission so he or she

understands the necessity of having two gears turning at the same speed before shifting. Once he or she understands the principle involved, additional training should be given to improve performance.

8. (C) One of the best methods for a trainee to learn how to do a new job is by doing it. At first, it should be done under the observation of the instructor. The instructor should allow the trainee to perform the first time without comment. The second time, the trainee should be shown and the various steps in the procedure should be explained. If the trainee makes a mistake, he or she should be stopped immediately before the wrong method becomes a habit.

9. (A) It has been said that experience is one of the best teachers. Whenever an instructor can tie a new idea to what a trainee has previously gained from experience, the better the chance that the new idea will be remembered. Instructors must be alert, however, to the fact that the experiences of those being taught vary. It is important to utilize several examples in the hope that the examples parallel the assorted experiences of the learners.

10. (A) Based on a scale of 100, seeing rates approximately 87% for effectiveness in learning. Hearing is number two, accounting for approximately 7% while the other senses account for 6%. It is unfortunate that most instructors depend almost 100% on the hearing sense, while providing instruction. For maximum learning effectiveness, an instructor should attempt to utilize all five of the senses.

11. (B) Generally, the combined thinking of a group is much more effective than the thinking of any one individual. A company officer who uses this approach in problem solving will probably see ideas developed which might not have manifested themselves if he or she tried to solve the problem alone, or worked with another individual on the solution. Not only are more ideas likely to develop from a training situation such as this, but the people involved are made to feel that they are a part of the organization when their ideas are solicited and used.

12. (C) The four-step method of training has proven to be extremely successful. The steps in proper sequence are: prepare, present, perform, follow-up.

13. (D) A conference as a training method is most effective when there is a problem to be solved. The conference leader should utilize the diversified backgrounds of the members present to gather as much input as possible into the solution to the problem. The participants should have some experience in the area under discussion and must be made to contribute as much as possible. The task of getting full participation rests with the conference leader.

14. (D) Skills cannot be acquired by reading a book or watching another person perform. In order to acquire a skill, a person must perform. Even after a person becomes skilled, it takes practice to maintain that skill. A good example is the constant need to

drill on hose lays in order not to become rusty. Some may consider the old saying, "practice makes perfect," nothing but a cliche, but it does contain a grain of truth.

15. (D) Permitting the trainee to correct himself or herself is the best of the correcting techniques. No one likes to make a mistake, and no one likes to be criticized for making one. Much of the unpleasantness of making a mistake can be eliminated if the trainee is able to correct his or her mistake. One method of achieving this is to compliment the trainee and then ask if he or she can think of anything that could have improved his or her performance. Of course, if he or she does not realize a mistake has been made, then the instructor should offer a suggestion for improving the performance.

16. (D) A person who is extremely knowledgeable in a given area will not necessarily make a good instructor. In fact, some people are so knowledgeable that they fail to see the subject from the learner's viewpoint. A good instructor is one who knows how to organize material into a logical method for presentation, knows how to motivate learners, uses audio and visual aides to assist in clarifying points, and one who can evaluate the learner's interest and response.

17. (B) If a man has the ability to learn, and is not performing properly, it is an indication that he probably needs training. Some supervisors regard mistakes as carelessness on the part of the workers. They should realize that they are the careless ones, as all supervisors are responsible for ensuring that their subordinates are properly trained.

18. (B) The old saying, "you can lead a horse to water, but you can't make him drink" has application in the field of training. You can conduct a training session, but you can't make the trainees learn. Learning only occurs when a person wants to learn. Consequently, the task of the instructor is to instill the will to learn into the trainees. One method to accomplish this is to show them why learning the material, or skill, is of importance to them. There are a number of motivating factors that will cause people to want to learn. Some of these are personal safety, promotion, or even job maintenance.

19. (D) A training program will succeed or fail depending upon the ability of the instructors more than on any other single factor. Consequently, a good training program is one with clearly defined objectives and instructors who both know their subject matter and are skilled at teaching.

20. (D) As mentioned in Question 5, training should result in change. The best method of measuring the results of a training program is to determine the changes made in the trainees. If the changes can be measured in performance, then the difference in performance prior to the training and after the training is a measure of the changes that took place during the training period.

Public and Community Relations

1. The relationship between the fire department and the press is like a "two-way street" because the press is not only a medium through which the department releases information to the public, but the press also

    (A) is interested in the promotion of the department's programs.
    (B) can teach the department good public relations.
    (C) makes the department aware of public opinion.
    (D) provides the basis for community cooperation with the department.

2. You arrive in your office at 11 a.m., having been on an inspection tour since 8 o'clock. A man has been waiting in your office for two hours. He is abusive because of his long wait, and accuses you of sleeping off a hangover at the taxpayers' expense. You should

    (A) say that you have been working all morning, and let him sit in the outer office a little longer until he cools off.
    (B) tell him you are too busy to see him and make an appointment for later in the day.
    (C) ignore his comments, courteously find out what his business is and take care of him in a perfunctory manner.
    (D) explain briefly that your duties sometimes take you out of your office and that an appointment would have prevented his inconvenience.

3. Of the following, the proper attitude for an administrative officer to adopt toward complaints from the public is that he or she should

    (A) not only accept complaints but should establish a regular procedure whereby they may be addressed.
    (B) avoid encouraging correspondence with the public on the subject of complaints.
    (C) remember that it is his or her duty to get a job done, not to act as a public relations officer.
    (D) recognize that complaints are rarely the basis for significant administrative action.

4. A fire department may expect the most severe public criticism

    (A) when asking for an increase in the budget.
    (B) when attempting to float a bond issue for new equipment.
    (C) after a fire which caused a large loss in property.
    (D) just after a budget increase has been granted.

5. As a company officer you receive a complaint from a citizen regarding the department's actions which you believe is unwarranted. The best action for you to take is to

    (A)   ignore the complaint.
    (B)   write a letter to the citizen informing him that his complaint is unsound.
    (C)   file the complaint and remember to talk to the citizen the next time you are in his area.
    (D)   give it immediate attention and investigate it thoroughly.

6.   When considering the relationship between business and governmental relations, it can best be said that

    (A)   the two are identical.
    (B)   there is no similarity between the two.
    (C)   there are many similarities between the two but there are some important differences.
    (D)   there is only one similarity between the two but it is important.

7.   Obtaining feedback from the public is an important part of the public relations effort of any public agency. Perhaps the most difficult people to reach are

    (A)   those from the lower income area.
    (B)   homeowners' groups.
    (C)   service clubs.
    (D)   religious organizations.

8.   Suppose you have been asked to answer a letter from a local board of trade requesting certain information. You find you cannot grant this request. Of the following ways of beginning your answering letter, the best way is to begin by

    (A)   quoting the laws or regulations which forbid the release of this information.
    (B)   stating that you are sorry that the request cannot be granted.
    (C)   explaining in detail the reasons for your decision.
    (D)   commending the organization for its service to the community.

9.   A fire officer should, as far as operations permit, answer any reasonable questions asked by the occupant of the premises involved in a fire. However, of the following, the subject about which he or she should be most guarded in comments is the

    (A)   time of the arrival of companies.
    (B)   methods used to control and extinguish the fire.
    (C)   specific cause of the fire.
    (D)   probable extent of the damage.

10.   In a governmental agency, the basic objective of being directly concerned with public relations should be to

    (A)   promote the most efficient administration of the agency.
    (B)   produce annual reports which will be acceptable to the public.
    (C)   increase the size of the agency.

(D) broaden the "scope and activities of" the agency.

11. The most effective time for a fire department to hold "open house" at the fire station is

(A) during the winter, when the fire department has the most calls.
(B) when firefighting is receiving general public attention as during Fire Prevention Week.
(C) during the summer months when children are not in school.
(D) immediately after a disastrous and widespread fire.

12. A local fire prevention committee can be very useful in reducing fire and life hazards within a community. It is best that the chairperson of such a committee be

(A) a person having a strong commercial interest in the city.
(B) the fire chief.
(C) a representative from the city manager's office.
(D) a public-spirited citizen having no direct commercial interest.

13. One of the biggest misconceptions regarding governmental public relations is that they consist of nothing but

(A) gathering public opinion.
(B) publicity.
(C) trying to influence the public.
(D) increasing department prestige.

14. The most important function of a good public relations job of a fire department should be in

(A) training personnel.
(B) developing an understanding of fire dangers on the part of the public.
(C) enacting new fire laws.
(D) recruiting technically trained firefighters.

15. When employees consistently engage in poor public relations practices, which of the following is most often the cause?

(A) disobedience of orders
(B) lack of emotional control
(C) bullheadedness
(D) poor supervision

16. The aspect of public relations which has been most neglected by government agencies, but which has been stressed recently in business and industry, is

(A) use of radio.
(B) public speaking.
(C) advertising.
(D) training employees in personal contacts with the public.

17. Citizens' support of a fire department program can best be enlisted by

  (A) minimizing the cost to the community.
  (B) advising citizens how backward the community is in its practices and why such a situation exists.
  (C) advising citizens that it is their civic duty to do all that can be done to support the program.
  (D) advise citizens how the program will be of benefit to them and their families.

18. Assume you have been requested by the chief to prepare for public distribution a statement dealing with a controversial matter. For you to present the department's point of view in a terse statement making no reference to any other matter is, in general,

  (A) undesirable; you should show all the statistical data you used; how you obtained the data; and how you arrived at the conclusions presented.
  (B) desirable; people will not read long statements.
  (C) undesirable; the statement should be developed from ideas and facts familiar to most readers.
  (D) desirable; the department's viewpoint should be made known in all controversial matters.

19. It is important for every governmental administrator to recognize that public relations is

  (A) a necessary evil.
  (B) an important and inseparable part of his job.
  (C) an end, in and of itself.
  (D) something that should be delegated to a responsible subordinate.

20. The primary purpose of a public relations program of an administrative organization should be to develop mutual understanding between the

  (A) public and those who benefit by organized service.
  (B) public and the organization.
  (C) organization and its affiliate organizations.
  (D) personnel of the organization and the management.

21. The public is most likely to judge personnel largely on the basis of their

  (A) experiences.
  (B) training and education.
  (C) civic-mindedness.
  (D) manner and appearance while on duty.

22. The main advantage of good sound public relations is to

  (A) build up a good feeling and understanding between the department and the public.
  (B) gain public support for wage increases and better working

conditions.
(C) increase public interest in building projects.
(D) attain a friendly and sympathetic press.

23. One of the primary reasons why lower ranking employees in municipal organizations need training in public relations while lower
ranking employees in industrial operations do not is that

(A) municipal agencies need the support of the public more.
(B) municipal agencies operate on tax-supported funds.
(C) more lower ranking employees of municipal agencies come
in contact with the public than do employees in industrial operations.
(D) municipal employees need the contacts to improve wages
and benefits.

24. One of the biggest basic problems faced by civil defense authorities in establishing an effective civil defense program has been
in

(A) coordination with military authorities.
(B) recruiting competent top leadership.
(C) overcoming public apathy.
(D) predicting the extent of vulnerability to enemy attack.

25. Public relations can be considered both a concept and a process.
Which of the following would be considered a portion of the process?

(A) informing the public.
(B) influencing public support.
(C) measuring the public's attitude.
(D) communicating ideas to the public.

26. Public relations has been defined as the aggregate of every effort
made to create and maintain good-will and to prevent the growth of
ill-will. This concept assumes particular importance with regard
to public agencies because

(A) public relations become satisfactory in inverse ratio to
the number of personnel employed in the public agency.
(B) legislators may react unfavorably to the public agency.
(C) they are much more dependent upon public good-will than
are commercial organizations.
(D) they are tax supported, and depend on the active and intelligent support of an informed public.

27. In dealing with the public it is helpful to know that generally
most people are more willing to do that for which they

(A) are not responsible.
(B) understand the reason.
(C) will be given a little assistance.
(D) must learn a new skill.

28. The individual responsible for public relations in a fire depart-
    ment must constantly remember that there is (are)

    (A) only one public.
    (B) two different publics.
    (C) three publics.
    (D) many publics.

29. One way to make written material more understandable to the public
    is to avoid big words. Another good way is to

    (A) repeat all important ideas.
    (B) include only one idea in each paragraph.
    (C) use short sentences.
    (D) clearly define all terms.

30. One of the most widely used definitions for public relations is
    that P (for performance) plus R (for reporting) equals PR (for pub-
    lic relations). This definition is

    (A) excellent because it covers every aspect of public rela-
         tions.
    (B) incomplete as it gives equal weight to both performance
         and reporting.
    (C) acceptable but incomplete because prestige is not in-
         cluded.
    (D) an oversimplification.

Answers

1. (C)   The two-way street mentioned in the question is that the department gives the press information and the press, in turn, provides information for the department. Whoever is responsible for public relations in a fire department should be alert to public opinion of the department as expressed through the press media. This information is extremely useful in conducting the department's public relations affairs.

2. (D)   First of all, keep your cool. You can't be two places at the same time. It is important for people to understand that the fire department is a business, and is operated in a business manner. Most people who want to see a particular person in a business organization will call to make sure the person will be in before making the trip to the office. Do not make excuses. Briefly explain to the person that your duties sometimes take you out of your office and that an appointment would have prevented the inconvenience.

3. (A)   The administrative officer should constantly be reminded that the public is his or her boss. Not only is it important to accept complaints from that boss, but he or she should establish a system of communications with the public, and a system whereby all complaints are satisfactorily handled. The administrative officer's job is to provide the service demanded by the public.

4. (C)   Criticism is generally based on a lack of understanding. Most people do not accept the fact that some things just go wrong. When things go wrong, someone is to blame. Whenever a fire resulting in a large loss of property occurs, the fire department should expect that a certain number of people will criticize the department's operations. The primary method of counteracting this criticism is to provide adequate information regarding the causes of the loss. Although a department can expect the most criticism after a large loss fire, it can also use the fire to push for more stringent building or fire codes, or additional equipment or manpower to improve the department's attack capabilities. It must be remembered, however, that the lack of these items must honestly have been a contributing factor to the large loss and should not merely be used as an excuse.

5. (D)   A complaint is a warning that a citizen is not happy with some part of the department's performance. Complaints should be given immediate attention and investigated thoroughly. What may seem trivial to a department member is important to the complainant. It is important that the complaint be handled courteously and conscientiously.

6. (C)   Business and governmental public relations have many similarities but there are some important differences. Business public relations is very similar to advertising, with the general objective of increasing sales and reaching new clients. Governmental public relations, on the other hand, are more concerned with improvement of service to the public by keeping the public informed and ascertaining their needs.

7. (A) Generally, those people from the lower income areas are difficult to reach. For the most part, they do not have organized leadership, and many times they have neither the verbal nor writing skills required to make their needs and desires known. Additionally, many of them distrust any governmental organization, or anyone who represents government.

8. (B) The best method of starting any letter to a citizen or board when it is necessary to deny a request is to simply state that the request cannot be granted. Any information regarding the reason the request cannot be granted should follow the opening statement.

9. (C) An officer should always remember that any fire could result in a court case. The officer should be guarded whenever a discussion arises regarding the cause of the fire. A number of cases have been lost in court because of the careless remark of an officer or firefighter regarding the probable cause of the fire. Some departments guard against this on company fire reports by having the company officer note the probable cause of the fire rather than the cause. Many subsequent investigations by trained arson members have shown that the initial estimate of the company officer regarding the fire cause was in error.

10. (A) The promotion of public relations is an important part of any governmental agency. The basic objective of the public relations program should be to promote the most efficient administration of the agency. Every administrator should make the function of public relations an accepted part of his or her responsibilities.

11. (B) Fire prevention week and open house "married" in the United States with the initiation and development of Fire Service Day--a Saturday during Fire Prevention Week when firehouses all over the country open their doors to the public. Fire departments develop exhibits and plan all kinds of firefighting displays for the public on this day, making it a national public relations affair.

12. (D) Local fire prevention committees can be extremely useful to fire officials in helping to reduce loss of life and property from fire. It is best that the committee represent a true cross-section of the community. Membership should include city officials, school officials, merchants, representatives from insurance groups, service clubs, homeowners' associations, the press, etc. It is best that the fire chief is not the chairperson. The head of the committee should be an enthusiastic citizen who does not have any commercial interest in the community.

13. (B) Public relations programs have little long-term value if the primary objective is publicity. This approach neglects the important aspects of communicating with the public, gathering public opinions and attitudes, and keeping the public informed of operational goals.

14. (B) Another phase of public relations in a fire department's operations is the "training of the public." Part of the public training is designed to provide an understanding and an appreciation of department operations, but the most important part relates to fire

prevention and public safety. The department is responsible for training people to understand the dangers of fire and how to protect themselves in the event of emergencies. This training is conducted through the various media, in the schools, and during home inspection programs.

15. (D) Company members generally reflect the attitude of the company commander. If the company commander has a positive attitude, the crew will probably have a positive attitude. However, if the company commander has a laissez-faire style of leadership, the crew will probably do the same, or maybe worse. When firefighters engage in poor public relations practices, it is probably due to poor supervision.

16. (D) For many years governmental administrators believed that public relations was a management function with which lower ranking employees should not concern themselves. Over time, administrators have come to realize that every employee is a public relations officer. The public judges an agency by the contacts made with its people and most contacts are with employees at the lower level of the hierarchy. It has therefore become a concern of most governmental agencies to train all employees in public contact in order to improve the image of the organization.

17. (D) In general, the public supports the fire service. However, whenever the department wishes to place a new program into operation, it is necessary for the department to educate the public regarding the need for the program. Just informing citizens of the need is not enough. To gain support, it is necessary that they understand how the program will directly benefit them and their families.

18. (D) Although it is important for the department to provide the public with information regarding the department's viewpoint in a situation such as this, it is more important that it be presented in a manner that the public will understand. The best way to accomplish this is to develop it from ideas and facts that are already familiar to the public being addressed. The principle involved is similar to one used in training. When teaching something new, tie it to something the trainee already knows.

19. (B) Public relations is an important and integral part of the job of every governmental administrator. It has a distinct impact on decision making and effective performance. Only through adequate information from the public can an administrator make decisions in terms of political and administrative feasibility.

20. (B) It is essential that a fire department in its public relations program develop mutual understanding between the department and the public it serves. The program should keep the public aware of department operations, fire prevention and safety programs while also providing avenues and being alert for feedback regarding the public's opinions of the department.

21. (D) People judge other people and things on what they see. Very few people have the opportunity to observe firefighters while they

are working at emergencies. Consequently, they judge the firefighters' effectiveness by their manners and appearance during fire
prevention exercises, while visiting an engine house, or at other
locations where they may encounter firefighters who are on duty.

22. (A) One of the main objectives of a public relations program is
to provide and maintain a two-way street between the organization
and the public. The purpose of this two-way street is to develop
and maintain mutual respect between the two, and to provide a system for keeping each other informed. The public must understand
and appreciate the operations of the department, and the department
must understand and appreciate the needs of the public.

23. (C) Nearly every municipal employee will normally have some direct
contact with the public. This is in direct contrast to the situation in large industrial organizations where most employees do not
come in contact with the public. Consequently, it is necessary
that employees in public organizations be trained in public relations in order to better achieve the goals of the organization.

24. (C) This apathy of the public regarding civil defense has increased
as far as the preparation for defense against a nuclear attack is
concerned. The apathy, at one time, was based on the belief that
"it won't happen to us," but has now changed to the attitude, "there's
little that can be done if it does happen." Trying to get the public to prepare for evacuation in the case of a nuclear attack has
become almost impossible.

25. (D) Public relations as a concept generally means informing the
public, influencing public support, and measuring public opinion.
Public relations as a process is concerned with communicating with
the public.

26. (D) The public is the boss of governmental organizations. In order to receive the cooperation and support of the boss, it is necessary that he or she be kept informed, that he or she knows and
understands the necessity of maintaining a particular size force,
and that employees perform effectively.

27. (B) The basis of fear is the unknown; people tend to resist or re-
ject things they do not understand. Distrust is built upon a lack
of understanding. But, on the other hand, most people are willing
to accept and support operations when they understand the reasons
for them.

28. (D) Public relations, in essence, means dealing with the public.
However, it is important to remember that there are many different
publics. For example, preparing material for use by children in a
fire prevention program is much different than preparing material
for the use of parents in a home escape plan. Material must be
prepared for the citizenry it will serve. In some communities, it
is even necessary to prepare public relations material in several
different languages. Although a fire department does not have as
many publics as other municipal departments, it should not assume
that all publics are the same.

29. (C) Material prepared for public consumption should be written so that it is easily understood by all of the people who will read it. Keep it simple. One of the best ways to do this is to avoid the use of big words and keep sentences short. Both of these factors should be considered when determining the reading level of written material.

30. (D) The definition is oversimplified. It neglects many dynamic factors such as opinions, attitudes, feedback, and the communications process, that are required for good public relations. Even the factor of the leadership of the organization is missing.

Safety

1. The first important step for the company commander to take in the establishment of a safety program at the company level is to

   (A) instruct all firefighters in safety procedures.
   (B) enforce the rules.
   (C) sell himself on the importance of safety.
   (D) solicit the help of the company members in the development of the safety program.

2. While giving job instructions on a new tool, a captain is asked by a member why the operation was not performed in a different manner. The method suggested included some actions that could cause injury. The captain answered that the method he was demonstrating was the correct one, that the suggested method was unsafe, and there was no point in discussing wrong methods. The captain's approach to the question was

   (A) proper, mainly because the members will not have a chance to pick up bad habits.
   (B) improper, mainly because the captain did not consider the possibility of modifying the suggestion to make it safer.
   (C) proper, mainly because speed of learning is most rapid when only one method is followed.
   (D) improper, mainly because hazards of incorrect methods can be avoided if they are known.

3. Of all the factors needed to achieve a good safety record, the one that is most important is undoubtedly the

   (A) condition of the equipment used by personnel.
   (B) support given to the program by the chief.
   (C) safety training provided by the department.
   (D) attitude of department personnel.

4. From a given electrical source, the extent of electrical shock received by a person who comes in contact with a live wire will depend upon

   (A) the path the current takes through the body.
   (B) the frequency of the voltage.
   (C) the duration of the contact.
   (D) a combination of the above three factors.

5. Of the following, the main reason for a fire officer to stress safety is to

   (A) develop a respect for danger in all members.
   (B) replace ignorance with practical knowledge leading to the elimination of unfounded fears.
   (C) establish a safety-conscious work atmosphere in which firefighters seek safer operating methods.
   (D) reduce the degree to which experienced members express impatience and contempt for "restrictive" safety practices.

6. The best method of making employees safety-conscious is through

   (A) the use of a safety training program.
   (B) the use of an accident review board.
   (C) the use of an accident committee.
   (D) disciplining employees for violating safety regula-
   tions.

7. There are several reasons for making an accident investigation.
Which of the following <u>is not</u> a legitimate reason?

   (A) determining who was at fault for the purpose of dis-
   ciplinary action
   (B) to improve the effective use of personnel
   (C) for determining hazardous conditions that caused or
   contributed to the accident
   (D) to improve the morale of the department

8. While taking up lines during freezing weather, a newly-appointed
captain sees a veteran firefighter freeing hose from ice by turning it back on itself and pulling it loose. Of the following, the
best procedure for the officer to follow in this situation is to

   (A) question the firefighter to find out whether he knows
   a better and safer way to perform this work.
   (B) assign the firefighter to other duties and assign another member to pick up the hose.
   (C) advise the member that if the hose is damaged the firefighter will be subject to charges for abusing department property.
   (D) order the firefighter to get the necessary equipment
   to do the job properly.

9. Firefighters should be taught the proper method of lifting. The
most common type of injury related to improper lifting techniques
is a

   (A) bruise.
   (B) sprain.
   (C) strain.
   (D) fracture.

10. The color of a turnout coat is an important factor in regard to
visibility. Tests have shown that the worst possible color for
visibility under both day and night conditions is

   (A) olive drab.
   (B) black.
   (C) yellow.
   (D) white.

Answers

1. (C) Although all the answers given are important to a safety program, the first of the important steps is that the company commander must be sold on the importance of safety. It isn't enough to
give lip-service to the program, as the subordinates will quickly
sense this. The company commander must be safety-conscious and
must develop this outlook in all of the company members.

2. (D) It is important in developing safety procedures for trainees
to learn what not to do as well as what to do. Most injuries are
caused by someone doing something wrong. Many times the injury is
caused by a person not realizing the possible results of performing incorrectly. It is therefore important that members be taught
the hazards of incorrect performance so that these procedures can
be avoided.

3. (D) The achievement of a good safety record is most likely due to
the attitude of the personnel toward safety than any other factor.
The firefighters must recognize the need for and the reasons behind
the safety program. Firefighters will be safety-conscious if their
attitude toward safety is positive; however, they may try to cir-
cumvent safety practices if their attitude is negative.

4. (D) All three of the listed factors have a direct bearing upon the
extent of the shock that will result from coming in contact with an
electrical wire. A current passing from one end of the body to the
other will more likely cause a more severe shock than one which
travels only a short distance through the body. Frequency affects
both the depth of the burn and the possibility of affecting the
heart, and the longer the contact with the wire, the greater the
extent of damage.

5. (C) Most plants having good safety records have achieved these
records by the establishment of a safety-conscious work atmosphere
within the plant. Workers must think safely as well as perform
safely to avoid accidents. Safety-conscious employees are able to
visualize the results of improper action before beginning an operation.

6. (A) A safety training program will instill in employees the need
to become safety-conscious. Of course, the first requisite is that
management be safety-minded. Employees generally are willing to
participate in such a program, if for no other reason than to learn
how to protect themselves from injury. They should be shown how
accidents are caused and what can be done to avoid them. This is
particularly true in firefighting situations as the profession itself is risky.

7. (A) The objective of accident investigations should be to determine the facts leading to or contributing to the accident. Accident investigations should not be made for the purpose of finding
out who was at fault so the person can be disciplined. The purpose
should be the elimination of hazards or unsafe practices so similar accidents will not occur in the future.

8. (D)   This situation requires that the company officer see that the job is done as quickly and safely as possible. He should order the firefighter to get the necessary equipment to do the job properly. The situation also waves a warning flag before the company officer that the firefighter needs further training in the use of equipment, and more indoctrination to become safety-conscious. This, however, should take place at a later time, under better conditions.

9. (C)   There are instances where all of the indicated injuries have been caused by someone using an improper lifting technique; however, back strains are the most common injury resulting from this faulty practice.

10. (B)   Turnout coats are available in all the colors listed. Black coats, however, were found to be undesirable for visibility. Although the lighter colored coats are more difficult to keep clean, they have proved to be more visible.

III

FIREFIGHTING TECHNIQUES AND STRATEGY

Principles and Procedures

Questions 1 through 15 are based upon information contained in the text _Fireground Tactics_ (Emanuel Fried, H. Marvin Ginn Corporation, Chicago, 1972). This information has been used with the expressed permission of the Ginn Corporation.

1. It has been said that the logical sequence to firefighting includes three phases. Which of the following is not one of these phases?

    (A) extinguish the fire
    (B) ventilate the fire
    (C) locate the fire
    (D) confine the fire

2. As company commander, you respond at 4:00 a.m. to a reported fire in a residential area. Upon arrival, you see fire shooting out of the first three floors of an abandoned four-story building. The fire is threatening an occupied four-story residential structure located nearby. The first line should go to the

    (A) threatened apartments.
    (B) fourth floor of the abandoned building.
    (C) first floor of the abandoned building.
    (D) second floor of the abandoned building.

3. A standard 2½-inch hand line can cover about how many square feet of area on a large-area interior fire in an industrial building?

    (A) 500
    (B) 1,000
    (C) 1,500
    (D) 2,000

4. It can best be said that the size-up of a fire begins

    (A) when the alarm is received.
    (B) when the fire building is first sighted.
    (C) with pre-fire planning.
    (D) when smoke or fire is first seen.

5. Four buildings of masonry construction are exposed to severe radiant heat at a large fire. Of the following, the one subjected to the greatest damage from the heat is constructed of

   (A) granite.
   (B) sandstone.
   (C) marble.
   (D) limestone.

6. The officer-in-charge of a major fire should operate from a

   (A) position in front of the fire.
   (B) roving position in order to constantly evaluate the exposure problem.
   (C) position inside the building, as close as possible to the main portion of the fire.
   (D) position to the rear of the fire building.

7. There are several common errors made by fire officers on the fireground. Which of the following is considered the major mistake?

   (A) failure to supply sprinklers early
   (B) failure to use self-contained breathing equipment
   (C) the officer-in-charge vacating his frontal command post to go into the fire building to survey the situation personally
   (D) underestimating the fire potential

8. A truck officer ventilating a building has a number of objectives in mind. Which of the following is <u>not</u> considered a primary object?

   (A) saving lives
   (B) reducing the extension of the fire
   (C) reducing water damage
   (D) making it possible for firefighting forces to move in

9. A firefighter is ordered to ventilate the roof of a burning building. He has a number of choices as to how to get to the roof. Of the following, which is most desirable?

   (A) up the interior stairway of the fire building
   (B) up the aerial ladder to the fire building
   (C) from the roof of an adjoining building of the same height
   (D) using the elevator of the fire building

10. Upon arrival at the fire building, the officer-in-charge notices that the building is heavily charged with smoke from bottom to top. From past experience, he should realize that the fire is probably

    (A) in the cellar.
    (B) on the first floor.
    (C) on the top floor.
    (D) on a floor somewhere near the midpoint between the first and top floors.

11. There are several techniques available for detecting concealed fires. Which, if any, of the following cannot be considered one of the techniques?

    (A) touch
    (B) sight
    (C) smell
    (D) all of the above

12. A fire is reported to be in the cockloft of a multistory building. As a firefighter you should know that the fire is probably located

    (A) on the roof.
    (B) in the attic.
    (C) immediately below the roof.
    (D) in a separate room on the roof.

13. Many structure faults are built into a structure on the drawing board. Probably the worst architectural fault is

    (A) unprotected vertical openings.
    (B) large open floor areas.
    (C) unprotected metal structural members.
    (D) alterations in building occupancy.

14. There are several ways to get hose lines to upper floors of buildings under construction. Of the following, which one is considered the best?

    (A) hand-stretching lines up the temporary inside stairs
    (B) taking lines up an aerial ladder
    (C) pulling the lines to upper floors by the use of hose
        rollers
    (D) using an elevated platform as a standpipe up to its
        limit, then extending lines from there to the fire
        by the use of a hose roller

15. Probably more often than in any other case, buildings are destroyed by fires that start in

    (A) attics.
    (B) storage rooms.
    (C) cellars.
    (D) kitchens.

16. Hosing a safe containing valuable records after it has been exposed to intense fire to cool it and gain access to its contents is generally a

    (A) good practice mainly because the lock mechanism will
        be protected from further heat damage.
    (B) poor practice mainly because the safe door may be
        warped by a sudden drop in temperature.
    (C) good practice mainly because further charring of the
        records will be prevented.

(D)   poor practice mainly because there may be flash ignition of the contents when the door is opened.

17.   Before fire departments took on the job of salvage, this function
was

(A)   unnecessary; in those days fires usually resulted in
total loss.
(B)   needed, but not done by anyone.
(C)   performed by public works departments.
(D)   performed by the fire insurance companies.

18.   Before overhauling is begun, it is most essential for the officer-
in-charge to

(A)   arrange for the electricity to be shut off.
(B)   determine if the building is structurally safe for work
parties.
(C)   make sure that all fires are extinguished.
(D)   clear the ground adjacent to the building so that arti-
cles may be thrown out of windows.

19.   "A tendency to call for help only after the force on hand has def-
initely proved inadequate is inexcusable." This point of view
is

(A)   correct; by the time help comes the fire may be extended beyond the control of the added force.
(B)   not correct; calling for help before it is needed
results in a needless expenditure.
(C)   correct; a chief should always request more help.
(D)   not correct; if this is carried through to its logical conclusion there will not be a sufficient number
of companies in the city.

20.   The first major concern of a fire officer in combating a chimney
fire in a dwelling should be to

(A)   check for a hidden structural fire in the walls and
attic.
(B)   place salvage covers to protect interiors from water
damage.
(C)   place ladders against chimney.
(D)   get streams in operation to cool the outside of the
chimney.

21.   Perhaps the most important factor of a building that should be considered by the commanding officer in sizing up a fire is the

(A)   occupancy.
(B)   construction.
(C)   height.
(D)   elevators and other shafts.

22.   Back draft is most likely to occur at a fire in a building charged
with fumes and smoke and in which the fire has been burning for

some time and is fairly well closed in. Of the following, the first sign of a probable back draft is that as soon as the doors are opened

(A) flames will be seen to build up gradually.
(B) no flames can be seen.
(C) air is felt to rush out of the building.
(D) the smoke seems to be drawn into the building sharply.

23. "Cover work" at a fire should start as soon as possible

(A) whether or not the fire has been extinguished, on the ground floor and working up to the fire floor.
(B) after the fire is under control, on the floor immediately below the fire floor.
(C) whether or not the fire has been extinguished, on the floor immediately below the fire.
(D) after the fire has been brought under control, on the ground floor and working up to the fire floor.

24. In salvage operations, the pike pole drain is used primarily in connection with a

(A) stairway.
(B) window.
(C) door.
(D) parapet wall.

25. There are several factors which may be considered in determining the need for more firefighters at a fire. Which of the following is least important?

(A) likely spread of the fire
(B) number of hose lines needed to subdue the fire
(C) nature and proximity of the exposures surrounding the fire
(D) possible hazard to firefighting forces

26. Large fires in important occupancies are of such complexity that

(A) no planned procedures are reliable.
(B) no plans or methods of attack can be formulated in advance.
(C) the problem must be considered in advance and methods of attack formulated.
(D) a committee of chief officers is necessary to direct firefighting at the fire.

27. Pier fires are the most difficult to extinguish once started. There are many factors that seriously affect the firefighting problems in these places, but the one outstanding feature of pier construction that allows rapid spread of fire is

(A) lack of sprinkler protection.
(B) their extensive areas.
(C) their isolation from other structures.

(D)   their proximity to waterway.

28.   When wetting agents are used to modify the properties of water used for extinguishing fires, their principle advantage is that

(A)   they may be used indiscriminately in fire extinguishers depending on a balanced chemical reaction.
(B)   greater penetration is attained by reducing the surface tension of water.
(C)   their biggest advantage is in situations requiring large volumes of water for cooling.
(D)   their low cost offsets any possibility of damage.

29.   A captain is the first to arrive at the scene of a building reported to be on fire. As he approaches the building, he sees grayish-yellow smoke emitting intermittently from cracks around the door. To the captain this will most likely indicate that

(A)   the gas furnace has exploded.
(B)   the residence is being fumigated with chlorine gas.
(C)   the fire department is the victim of a hoax.
(D)   conditions are favorable to a backdraft.

30.   When fire spreads to the stairs, elevators, and corridors of an old non-fireproof hotel, the greatest life hazard is usually found

(A)   in the collapse of interior stairs at lower floor levels.
(B)   on the top floor.
(C)   in places of public assembly above the ground floor.
(D)   in cellars and subcellars.

31.   After use, an exterior standpipe system should be <u>first</u>

(A)   drained at the second floor level.
(B)   flushed with an approved cleanser.
(C)   tested for leaks.
(D)   reported to the building engineer or owner.

32.   In making a stairway drain that requires two salvage covers, the lower salvage cover is thrown from the

(A)   bottom of the stairs to the center and the upper cover from the center to the top of the stairs.
(B)   bottom of the stairs to the center and the upper cover from the top to the center of the stairs.
(C)   center to the bottom of the stairs and the upper cover from the top to the center of the stairs.
(D)   center to the bottom of the stairs and the upper cover from the center to the top of the stairs.

33.   "Because no two fires are alike, it is impossible to lay down general plans for firefighting operations." This viewpoint is unacceptable primarily because

(A)   elements of similarity are sufficient to establish tactics and strategy applicable to a variety of situations.

(B) the variety of techniques and methods available are conducive to training at all levels of command.

(C) proper utilization of the forces at hand are independent of the variable characteristics of fire situations.

(D) fighting major fires involves essentially an extension of the problems involved in minor fires.

34. "It is clear that the complexity of the problems involved in fighting large fires is such that officers cannot be expected most efficiently to solve them alone on the fire grounds." Which of the following is the most important implication of this statement?

(A) Large fires are seldom frequent enough to enable a chief officer to develop real proficiency in handling the complex major fire.

(B) The difficulty of controlling officers and firefighters of varying ability at major fires precludes a concentrated balanced attack.

(C) Emergency action, usually necessary at major fires, usually prevents the application of preplanned attack.

(D) Prior knowledge of the hazard and its problems is essential for most effective action when the fire occurs.

35. Generally, when attacking a backstage fire in a large theater during a performance, which of the following actions would be most questionable from a firefighting point of view?

(A) closing the asbestos curtain
(B) opening the ventilators above the stage
(C) bringing in the first line backstage
(D) bringing in lines onto the stage from both sides

36. Exclusion of air is one method of extinguishing a fire. However, as a practical matter, use of this method for fire extinguishment in common combustibles requires only that atmospheric oxygen be reduced from its normal concentration to a maximum of which of the following percentages?

(A) 15%
(B) 25%
(C) 35%
(D) 45%

37. When attacking a fire with fog, the nozzleman should

(A) advance slowly on the fire while moving the nozzle with a rotary motion.
(B) stand still while moving the nozzle from side to side.
(C) advance on the fire as rapidly as is possible while moving the nozzle from side to side.
(D) stand still while moving the nozzle in a rotary motion.

38. You arrive first at an automobile fire. A short circuit caused
     a fire under the hood which had spread to the dashboard. When you
     arrive the wires are smoldering. Of the following acts, which
     should be performed first?

     (A)   disconnect the battery terminal connections
     (B)   cool the hood by application of water from a 1½-inch
           hose
     (C)   apply ashes, sand, or earth to smother the fire
     (D)   cut the ignition wiring using a linoleum knife

39. When a fire is visible from the street at the front windows of a
     building, it is very frequently reasonable to assume that the fire
     has originated close to the windows and that a very small portion
     of the building or floor is involved. In such instances, the most
     efficient attack is from the

     (A)   outside of the building primarily because it is a
           more direct method of attack.
     (B)   inside of the building primarily because streams
           can be pointed directly at the fire.
     (C)   outside of the building primarily because heavier
           streams can be brought to bear on the fire.
     (D)   inside of the building primarily because the hazards
           of an outside attack in varying weather conditions
           can be avoided.

40. Suppose a firefighter wishes to force a locked window open so he
     can enter a burning building. It is the usual double-hung kind
     of window which slides up and down. The best action for him to
     take is to insert the blade of an axe

     (A)   between the upper and lower halves of the window and
           twist sharply.
     (B)   beneath the base of the lower sash and pry upward.
     (C)   between the window and the window frame on the right
           side and pry inward.
     (D)   between the upper window pane and sash and twist gently.

Answers

1. **(B)** The three phases in the logical sequence to firefighting are:

   <u>Phase One</u>--Locate the fire. (Not always an easy job; too many firefighters still use water on smoke.)
   <u>Phase Two</u>--Confine the fire (head it off and surround it.)
   <u>Phase Three</u>--Extinguish the fire.

2. **(A)** The first line should go to the threatened apartments. The second line should cover the floor above the threatened apartments, if necessary.

3. **(B)** Industrial fires of any consequence call for all interior lines to be 2½ inches. This requires two firefighters at the nozzle, a third lining up behind, and a fourth pulling surplus hose. If you are "making" a stairway you will need a fifth firefighter. On large-area interior fires, each standard 2½-inch hand line can cover about 1,000 square feet of area.

4. **(C)** The size-up should begin with pre-fire planning. This requires a detailed inspection, not only of the physical plant but of the surrounding areas as well. This will provide valuable insight on the kind of construction, the location of stairs and exits, whether the building has sprinklers or standpipes, the kind of occupancy (hazardous, explosive, chemicals), fire walls and fire doors, exposures both internal and external, water supplies, etc.

5. **(D)** Limestone is subjected to greater damage from heat than the other three. At about 800°F, limestone decomposes into lime. It crumbles and flakes at high temperatures and is subjected to expansion and contraction stresses.

6. **(A)** He should be in front of the fire. He should be assigned an aide who has been trained in his strategy and who needs little instruction from the chief at the fire. The aide should do the chief's legwork and help coordinate incoming units and equipment. The chief should use competent officers to survey the fire for exposure threats and report the situation to him.

7. **(D)** All the factors listed are common errors made on the fireground; however, the usual major mistake is underestimating the fire potential. Some additional common errors are

   1. failure to head off the fire initially
   2. failure to realize that exhausted men will need relief
   3. failure to call mutual aid soon enough

8. **(C)** Effective ventilation can accomplish three desirable objectives, as follows:

   1. Save lives. Ventilation will draw fire, heat, and smoke away from trapped people.

2. Reduce the extension of fire through lateral spread or the mushrooming effect.
3. Make it possible for firefighting forces to move in. Firefighters can reach the seat of the fire more quickly and with the least physical punishment.

9. (C) Getting to the roof from a roof of an adjoining building of the same height is the preferred method. It is also the safest. A good example of the hazards of using the stairway of the fire building is as follows: As you are trying to force the roof door open from the stairway side, another man is sent up the aerial to the roof to open up. He forces the door from the roof side. Thus fire from below now is pulled up the shaft instantly and traps you right in its path.

10. (A) The following are some basics to remember:

1. Smoke pushing out of upper floors is no guarantee that the fire has extended to them.
2. As a rule, smoke from an upper floor fire will not fill the lower floors.
3. When the building is heavily charged with smoke from bottom to top, the chances are you have a cellar fire.

11. (D) Touch--Feel the suspected area with your bare hand.
Sight--Look for seepage of smoke from suspect areas or above them.
Smell--Smoke from a live fire will have a different smell than trapped or "dead" smoke. This means that seepage of smoke from a suspected area will not always indicate fire. The fire may have seeped up from a fire below and been slow to dissipate via cracks, crevices, etc. However, if in doubt, find the source of smoke by opening walls, ceilings, etc.

12. (C) Cockloft is a term commonly used in fire department language to denote a blind space at the top of a structure in a flat or trussroofed building. This space exists between the finished ceiling you can see on the top floor and the underside of the roof sheathing.

13. (A) Probably the worst architectural fault, and certainly the one responsible for the greatest loss of life in buildings, is the unprotected vertical opening. This refers to open stair shafts, elevators, light and vent shafts, pipe chases, etc. This basic fault allows a fire on the lower floor to expose all floors above. The combustion products sweep up and involve all areas of the building. Not only are people overcome, but the normal means of exit from the building are cut off.

14. (D) Perhaps the best method is to use an elevating platform as a standpipe up to its limit (about the eighth floor). From here the firefighters can carry donuts to the floor below the fire and lower them over the hose roller by rope to the basket of the elevating platform.

15. (C) Probably more buildings are destroyed by fires that started in cellars than in any other place in a building. There are a number of reasons. Some of the important reasons are as follows:

     1. As a rule, the combustion process in cellars is slow.
     2. Discovery of such fires usually is delayed.
     3. Ventilation is limited.
     4. Firefighters going down into a cellar fire are in essence moving down into a chimney.
     5. Cellar fires usually require breathing masks.
     6. The application of water to the seat of the fire is frequently inefficient.
     7. It is difficult to determine whether the fire is moving up beyond the reach of the hose streams.
     8. The technique of handling such fires is not well understood.
     9. A cellar fire exposes all floors above the cellar to the action of the combustion products which fill the upper floors rapidly.
     10. The material stored in most cellars provides sufficient fuel for a hot fire.

16. (C) A hot metal container, such as a safe, will hold the heat for a considerable time. Consequently, the material inside will continue to char due to the heat from the container. Absorbing the heat by hosing the container will prevent further charring and may result in saving contents that otherwise might be lost. It will take some time to absorb all the heat in the container by cooling it with water.

17. (D) Before fire departments took on the job of salvage, this function was performed by fire insurance companies. Each company had its own salvage squad. Companies' fire marks were placed over the doors of those buildings they insured. If an insurance company salvage squad responded to a fire and found that the mark over the door was not theirs, they left the scene of the fire. If the building was not insured, no salvage work was done.

18. (B) Before overhauling is started, it is essential that the building be examined to insure that it is structurally safe for firefighters to work inside. Particular attention should be given to floors. Floors can be weakened by the tremendous amount of water sometimes used in fire extinguishment, making them unsafe. Consequently, all water should be drained from the floors before overhaul operations commence. It is unwise to allow firefighters to work below a floor housing heavy machinery if there is any question as to the stability of the floor.

19. (A) A common error in firefighting procedure made by many command officers is to chase the fire. There is every excuse for over-estimating the force required to control a fire, but no excuse for underestimating the help needed. When making the initial call for help, it is not only necessary to size-up the fire in its present state, but to evaluate where it will be and what forces will be required to contain and extinguish it by the time help arrives. Even then, it is always best to ask for at least one engine

company more than is deemed necessary.

20. (A) One of the first major concerns of a fire officer at any fire not involving life is to insure that the fire will not spread. With a fire in a chimney, there is always the possibility that the fire could extend into the walls or the attic. These locations should be checked as soon as possible for further extensions.

21. (A) Knowing the occupancy of a building provides the officer-in-charge with two important facts: the problem of life hazard and what can burn. The saving of life should always be given first priority, while knowing what can burn provides the officer with some idea of what to expect in regard to fire intensity and rapidity of fire spread.

22. (D) This is where the word backdraft originated. Instead of smoke coming out of the building, it goes back in. When firefighters see this sign, they should realize that smoke being drawn in will soon reverse itself with an explosive effect. Firefighters should immediately take whatever steps are necessary to protect themselves from the explosive effect. Time may permit the individual to do no more than drop to the floor, covered with a turnout coat or whatever is available.

23. (C) The key to this question is "as soon as possible." Covering furniture and other perishables below the fire should always be started as soon as possible. Normally, covers are spread while the fire is in progress. It does not take long for water to work its way through the floor where the fire is, dripping onto the contents of the floor below the fire. Salvage work should start on the floor below the fire and be continued downward.

24. (B) A pike pole drain is one where pike poles are extended from a stepladder to a window with covers thrown over the poles. Another pike pole is used to poke a hole in the ceiling above the drain. Water from the floor above is thus directed into the drain and out the window.

25. (D) The possible hazard to firefighting forces has nothing to do with the number of additional firefighters needed at the fire. The primary factors used in determining the additional forces required are: 1) how many additional hose lines or heavy stream appliances will be required to keep the fire from spreading and to extinguish it; 2) how many additional men will be needed for rescue and salvage work, or for additional factors such as ventilation.

26. (C) Fires in large occupancies can become extremely complex. There is no excuse for not preplanning operations in these target hazards. The more knowledge officers and firefighters have about a particular occupancy, the more effective firefighting procedures will be in the event a fire does occur. Drills should be held with advanced plans for forcible entry procedures, ventilation, rescue, control of utilities, etc., in the event that fires occur in various portions of the building. Although fires can be tricky, the probable path of fire travel can usually be estimated quite accurately before the fire occurs.

27. (B) Unfortunately, piers and wharfs extend for considerable distances without fire stops. This, coupled with the accumulation of oil on pilings and a normal air movement beneath these structures makes fire spread rapidly once ignition takes place. The construction of the piers and wharfs makes it difficult to cut openings from the top to set up fire curtains. One of the most effective ways of stopping this spread is to place scuba divers in the water with lines in advance of the fire spread and set up a water curtain. Divers can then make a direct attack on the fire with additional lines.

28. (B) Wetting agents are additives that are used to lower the surface tension of water. Their most effective use is on fires that have burrowed into a mass of combustibles such as cotton.

29. (D) These are the signs of a potential backdraft. This means, if possible, that top ventilation should be provided before the room is entered. If impossible to top ventilate, the next best procedure is to cut a small hole into the area and insert a spray nozzle. Turn the nozzle to spray and work it in a revolving motion, directing the stream to the ceiling level. While doing so, make sure firefighters are in a position to protect themselves in the event that backdraft develops; however, this should not occur as long as room tightness is maintained.

30. (B) Of the four methods for heat to travel from one location to another (direct flame contact, radiation, conduction, convection), convection poses the biggest problem in multistory buildings with unprotected vertical openings. Once heat and smoke reach one of these vertical openings, they travel quickly upward to the top floor. This chimney effect results in a considerable loss of life from both smoke and fire on the top floor.

31. (A) When a standpipe is loaded with water, the pressure at the clapper valve at the inlet created by the head is excessive. It is therefore difficult to force open the clapper valve. The best method of draining the standpipe is to connect a 2½-inch line to the outlet on the second floor and extend the line to the ground. The line should not be equipped with a nozzle. The outlet on the top floor should then be opened. When this has been completed, the outlet on the second floor should be opened and the standpipe bled to this level. When all the water has been drained from the standpipe to the second floor level, the pressure on the clapper valve will have been reduced to approximately 5 psi, making it fairly easy to force open the clapper valve to complete the draining of the standpipe.

32. (B) In making a stairway drain which requires the use of two salvage covers, the lower cover is thrown from the bottom of the stairs to the center while the second cover is thrown from the top towards the center. The top cover should overlap the bottom cover, preferably by at least one foot. The top cover should then be placed under the lip of the stairs at the top and held in place by a slat or similar object. Arrangements must be made to control the water when it reaches the bottom of the stairway.

33. (A) Although no two fires are exactly alike, all fires are somewhat similar. Consequently, basic principles of tactics and strategy can be developed with actual procedures modified at fires to fit the individual problems encountered.

34. (D) There is little doubt that pre-fire planning increases the chance of an effective operation when a fire occurs. The more a fire officer knows about his enemy, and what it will probably do, the more effective his operation will be when the battle takes place. (See question 26).

35. (C) One of the first acts of an officer when working on a fire backstage in a large theater is to insure that the asbestos curtain is dropped. This will normally confine the fire to the stage area. The ventilators above the stage should be opened next. To protect against the possibility that the asbestos curtain has not dropped fully, the first lines should be brought into the auditorium to protect against any spread into this area. Lines can then be brought onto the stage from the side entrances to work on the fire.

36. (A) Several figures concern the place where fire will be extinguished by oxygen reduction. The air normally contains 21% oxygen. When the oxygen concentration reaches 15% or 16%, the fire will normally go out. Some authorities state that complete extinguishment takes place at about 13%.

    It is interesting to note that these same figures apply to human beings. Life will cease when the oxygen concentration falls below 15% or 16%. Consequently, if there is not sufficient oxygen for the fire to burn, there is not sufficient oxygen to maintain life.

37. (A) Moving a nozzle in a rotary motion when applying fog is the most effective method of reducing the heat. Advancing slowly allows adequate time to reduce the heat, making it safe to proceed. If the fog is applied properly, there should be very little water runoff.

38. (A) A fire in the electrical system in an automobile is not much different from a fire in the electrical system in a structure. One of the first thoughts should be to disconnect the power. This can be done in an automobile by disconnecting the battery terminal connections.

39. (B) Normally, the best method of extinguishing a fire is to play streams directly on the fire. When the fire is inside a building, this means it is necessary to go inside to get directly at it. Approach the fire from an uninvolved area and push the fire toward the area that has already burned.

40. (B) Generally, these windows are secured shut by a catch located between the upper and lower window. The prying action of the axe will force out the screws holding this catch and permit the window to be opened without breaking the glass.

Shipboard Fires

Questions in this section are based upon information contained in the text <u>Marine Fire Prevention, Firefighting and Fire Safety</u> (Robert J. Brady Co., Bowie, Maryland). This information has been used with the express permission of the Robert J. Brady Company.

1. The ideal location for the staging area for a shipboard fire would be

    (A) the mess room.
    (B) the captain's cabin.
    (C) on an open deck, windward of the fire.
    (D) on an open deck, leeward of the fire.

2. An indirect attack would most likely be used when fighting a fire aboard ship when the fire is located in

    (A) the forecastle.
    (B) a cabin on the main deck.
    (C) the galley.
    (D) the lower portions of the vessel.

3. Which of the following statements regarding fires in ship passageways is most nearly correct?

    (A) The fire should always be attacked from the windward side.
    (B) The fire should always be attacked from the leeward side.
    (C) The indirect attack should always be used.
    (D) The fire must never be attacked from opposite directions.

4. A fire in an engine room has been knocked down by saturating the area with CO_2. At least how much time should elapse before entry is attempted into the engine room?

    (A) one-half hour
    (B) one hour
    (C) two hours
    (D) three hours

5. The kind of fire generally expected in a boatswain's locker is a

    (A) Class A fire.
    (B) Class B fire.
    (C) Class C fire.
    (D) Class D fire.

6. The receiver for the fire detection system aboard ship is normally found in

    (A) the engine room.
    (B) the forecastle.
    (C) the pilot house.
    (D) a separate compartment adjacent to the galley.

7. Of the following fires, which could be encountered in the galley of a ship, the kind of fire that would normally cause the most concern is a fire in the

   (A) deep-fat fryers.
   (B) plenum.
   (C) ovens.
   (D) frying griddles.

8. In vertical ventilating a fire under ideal conditions, the gases are released at a point directly above the fire as the extinguishing agent is brought to bear on the fire. This ideal vertical ventilation of fires aboard ship

   (A) can be achieved in engine room fires.
   (B) can be achieved in hold fires.
   (C) can be achieved in cabin fires.
   (D) is just about impossible to achieve.

9. To protect the exposures on a fire aboard ship, the fire must be surrounded on

   (A) three sides.
   (B) four sides.
   (C) five sides.
   (D) six sides.

10. A fire aboard ship may be considered to be under control when

   (A) the extinguishing agent is being applied to the seat of the fire.
   (B) the main body of the fire has been darkened.
   (C) all possible routes of fire extension have been examined or protected.
   (D) all of the above have been accomplished and a preliminary search for victims has been completed.

Answers

1. (C) The staging area should be established in a smokefree area, as near as possible to the fire area. An open deck location, windward of the fire, would be ideal. However, if the fire is deep within the ship, the staging area should be located below deck. A location near a ship's telephone, if feasible, would be helpful in establishing communication links. However, the staging area should not be located where it might be endangered by the spread of fire.

2. (D) An indirect attack is employed when it is impossible for firefighters to reach the seat of the fire. Generally, this is the case when the fire is in the lower portions of the vessel. The success of an indirect attack depends on complete containment of the fire. All possible avenues of fire travel must be cut off by closing doors and hatches and shutting down ventilation systems. The attack is then made from a remote location.

3. (D) A fire in a passageway must never be attacked from opposite directions. If it is, the hoselines will push flames, heat and smoke directly at the other hose team.

4. (B) A CO_2-saturated area should always be re-entered with caution. Although there are no hard-and-fast rules concerning re-entry, many factors must be considered. How hot was the fire? If oxygen is allowed to reach the fire area, will metal in that area be hot enough to cause reignition? Is it essential that the engine room be restored as fast as possible because of heavy seas? Or are the seas calm and without navigational hazards, so that entry may be delayed? The engine room should not be entered for at least an hour, primarily to allow the heat to dissipate. The injection of CO_2 into a sealed area will extinguish a flammable liquid fire almost immediately. However, since CO_2 has no cooling effect, metal surfaces remain hot. It is a no-win situation. The fire is out, but the threat of reignition makes the area dangerous.

5. (A) Access to the boatswain's locker is generally through a deck hatch opening in the forward section of the vessel. The hatch leads to the boatswain's stores, other storage areas, and the chain locker. The only access to the area is down a ladder and through a passageway. The area is used primarily for the storage of Class A materials.

6. (C) Fire detection systems on board a ship are so arranged that in case of a fire, both a visible and audible alarm is received in the pilothouse or fire control station (normally the bridge). For vessels of over 150 feet in length there also should be an audible alarm in the engine room. The receiving equipment (or consoles) indicate both the occurrence of a fire and its location aboard the ship. Consoles are located on the bridge and in the CO_2 room. The CO_2 room is the space that contains the fire extinguishing mechanisms. Only a bell is required in the engine room to alert the engineer to an emergency outside the machinery space.

7. (B) Three areas within the galley are especially subject to fire, as follows:

    1. The cooking area, including the frying griddles, boilers, deep-fat fryers, and ovens.
    2. The area immediately behind the filter screens, called the plenum.
    3. The duct system that vents heated gases.

    Fires in the cooking area can be serious. However, since they are out in the open, they usually can be extinguished completely. Fires in the plenum and the duct system are of most concern. Even after such fires are apparently extinguished, there may still be some fire hidden from view, or the fire may have extended out of the duct into nearby compartments. For this reason, automatic fire extinguishing systems should be installed to protect all parts of the range, plenum, and ducts.

8. (D) The smoke and hot gases generated by the fire should be vented to the outside air if possible. As a fire intensifies, the combustion gases become superheated. If they are ignited, they will spread the fire very quickly. In the ideal situation, the gases are released at a point directly above the fire, as the extinguishing agent is brought to bear on the fire. This ideal vertical ventilation is just about impossible to achieve aboard ship, since there is rarely a direct upward route from the fire to the outside. In most instances, at least some horizontal ventilation is required.

9. (D) Protecting the exposures means preventing the fire from extending beyond the space in which it originated. If this can be accomplished, the fire can usually be controlled and extinguished without extensive damage. To protect exposures, the fire aboard ship must virtually be surrounded on six sides; firefighters with hoselines or portable extinguishers must be positioned to cover the flanks and the spaces above and below the fire.

10. (D) A fire may be considered to be under control when:

    1. The extinguishing agent is being applied to the seat of the fire; i.e., streams from initial lines (and backup lines if they are required) have been able to penetrate to the seat of the fire and have effectively begun to cool it down. At this point, men with shovels should be able to turn over burned material to expose hidden fire.
    2. The main body of fire has been darkened. At this point, the fire cannot generate enough heat to involve nearby combustible materials.
    3. All possible routes of fire extension have been examined or protected. This is, basically, a combination of the exposure protection and overhaul procedures.
    4. A preliminary search for victims has been completed. The preliminary search should be conducted

at the same time as the fire attack, ventilation, and exposure protection procedures, if possible. As soon as the fire is under control, a second and more comprehensive search should be undertaken. Areas that were charged with smoke and heat must be closely examined. Searchers must look in closets, under beds, behind furniture and drapes, and under blankets. An unconscious person must be removed to fresh air immediately. If the person is not breathing, rescue breathing must be started immediately.

Hazardous Materials Fires and Emergencies

1. When companies respond to an alarm for a building where fumigation with hydrocyanic acid gas is being conducted, it is imperative that thorough and complete ventilation be effected immediately, primarily because in the concentration usually used

   (A) rapid corrosion of protective devices occurs.
   (B) application of water or foam is of limited effectiveness.
   (C) the gas is explosive even at room temperature.
   (D) the gas is absorbed through unprotected skin.

2. A kind of fire for which steam as an extinguishing agent is least effective is one involving

   (A) ammonium nitrate.
   (B) gasoline storage.
   (C) lumber drying.
   (D) rubber coating processes.

3. Water would be the most desirable extinguishing agent for a fire involving which of the following?

   (A) acetic acid
   (B) potassium cyanide
   (C) sodium
   (D) zinc powder

4. Suppose that at a fire a large tank containing hydrochloric acid collapses causing a possible explosion hazard by the acid on metal materials. Which of the following is the most desirable neutralizing agent to use on the acid?

   (A) chemical foam
   (B) carbon dioxide
   (C) soda ash
   (D) rock salt

5. Water is the most desirable extinguishing agent for hazardous chemical fires involving which of the following?

   (A) sulfur
   (B) potassium cyanide
   (C) sulfuric acid
   (D) zinc powder

6. Which one of the following fires can never be extinguished by smothering?

   (A) magnesium
   (B) acetone
   (C) sulfuric acid
   (D) nitro-cellulose film

7. The primary hazard of fighting fires involving the organic phosphate insecticide "Parathion" is the danger of

    (A) explosion.
    (B) extremely high temperatures.
    (C) flashback.
    (D) poisoning.

8. Water may safely be used as the extinguishing agent at a fire involving which one of the following chemicals or products?

    (A) calcium carbide
    (B) hydrochloric acid
    (C) sodium peroxide
    (D) molten magnesium

9. If a fire in a warehouse is about to involve many barrels of bleaching powder, the greatest danger would be from

    (A) an explosion.
    (B) a fast-burning flare-up that could not be controlled.
    (C) a toxic and irritant gas.
    (D) the impossibility of using water on the fire.

10. The extinguishing agent for a fire in potassium should be

    (A) water in large quantities.
    (B) water, in fog form, from a sheltered location.
    (C) carbon tetrachloride extinguisher.
    (D) sand or earth.

Answers

1. **(D)** Hydrocyanic acid gas is highly toxic. Exposure to a 0.3% concentration is fatal. This gas has been known to penetrate thick concrete walls, killing workers on the opposite side. At buildings where it is used as a fumigant, it is extremely important that the building be completely ventilated before any firefighter enters the premises. The gas can be lethal by absorption through unprotected skin.

2. **(A)** Ammonium nitrate is an oxidizing agent that should be considered an explosive under certain conditions. It is most likely to explode when subjected to heat, confinement, and pressure. Steam applied in a confined place would provide extra heat and eventually result in a pressure buildup. Fires in the early stages should be fought with large quantities of water. If a fire in a large quantity of the material is encountered in an advanced stage, unmanned heavy streams should be directed on the material with efforts to protect the exposures. Firefighters should be withdrawn to a safe distance. Breathing apparatus should be worn as the fumes from the fire may be toxic.

3. **(A)** The effect of using water on each of these materials is as follows:

    Acetic acid--Water in spray or fog form can be used.
    Potassium cyanide--Hydrogen cyanide gas (a deadly gas)
       will be released if this material comes in contact
       with moisture or acids.
    Sodium--This substance reacts violently to water when
       burning.
    Zinc powder--This material should be smothered with a
       suitable dry powder.

4. **(C)** Hydrochloric acid itself is not combustible but will produce hydrogen when it comes in contact with some metals. The result is an explosive mixture. If a spill of hydrochloric acid is encountered, firefighters should wear full protective equipment and neutralize the acid with soda ash or slaked lime.

5. **(A)** Burning sulfur gives off sulfur dioxide, a toxic gas. Fires involving this material should be fought with water spray. Firefighters should wear breathing apparatus.

    Potassium cyanide--Although noncombustible, hydrogen
       cyanide gas will be released when this material
       comes in contact with moisture or acids.
    Zinc powder--This forms an explosive mixture with air.
       Dry chemical should be used for fires involving
       this material.
    Sulfuric acid--Although noncombustible, this acid
       causes severe burns in contact with the skin. Heat
       is formed when this material comes in contact with
       water.

6. **(D)** Burning nitro-cellulose film gives off an extremely toxic gas.

Firefighters should always use breathing apparatus when this material is involved. Caution should be used as the material may explode under fire conditions. The best method of fighting fires in this material is to use large quantities of water spray to cool the material. Nitro-cellulose film fires cannot be extinguished by smothering with water as it is an oxidizer and will supply its own oxygen.

7. (D) Parathion is an extremely poisonous material. It can be fatal by skin contact, inhalation, or ingestion. Firefighters should always wear full protective equipment when working at emergencies involving this material. Fires containing this material should be fought with a water spray.

8. (B) The effect of using water on each of the chemicals is as follows:

> Calcium carbide--Acetylene gas forms in contact with water or moisture.
> Hydrochloric acid--This acid can be flushed away with water. Use soda ash or slaked lime to neutralize, if possible.
> Sodium peroxide--This material reacts violently with water. An explosion may occur.
> Molten Magnesium--Since it reacts violently with water, fires involving this substance should be extinguished by smothering with dry graphite or other suitable dry powder.

9. (C) The technical name for bleaching powder is calcium hypochlorite. The material is irritating to the skin, eyes, and respiratory tract. It is also poisonous if swallowed. Firefighters should wear full protective equipment when fighting fires involving this material. Fires should be fought using spray streams.

10. (D) Potassium is a combustible metal. Fires involving this material should be treated similarly to those involving magnesium. The material will reach violently with water when burning. It is best to use an appropriate dry powder. Of the choices given, however, sand or earth would be the most practical because use of any of the other three could cause a violent reaction.

Wildfires

Questions in this section are based upon information contained in the text <u>Wildfires - Prevention and Control</u> (Harry P. Gaylor, Robert J. Brady Company, 1974). This information has been used with the express permission of the Robert J. Brady Company.

1. Which of the following is probably the most universally used fire tool for wildfires?

   (A) round point shovel
   (B) square point shovel
   (C) council rake
   (D) brush hook

2. Which of the following methods of heat transfer is <u>least</u> related to wildfires?

   (A) radiation
   (B) direct flame contact
   (C) convection
   (D) conduction

3. Of the following weather elements that affect wildfires, which one is <u>least</u> predictable?

   (A) temperature
   (B) wind
   (C) relative humidity
   (D) barometric pressure

4. A foehn wind

   (A) blows from the ocean to the shore during the day.
   (B) blows from the shore to the ocean during the night.
   (C) flows from a high-pressure area on the windward side of mountains to a low-pressure area on the leeward side.
   (D) flows up a slope during the day because of surface heating and down a slope at night because of surface cooling.

5. Of the following, which fuel would ignite the most readily from burning embers?

   (A) moss and lichens in treetops
   (B) rotten wood, either on the ground, on logs, or in snags
   (C) slash, particularly when it is compacted in tight places
   (D) duff, peat, dried muck

6. Backfiring at wildfires should usually be used

   (A) as early as possible.
   (B) only when the winds have died down.

(C)   from controlled ridges only.
(D)   only as a last resort.

7.   The point and cutoff technique of fighting wildfires is most closely associated with

(A)   a flank attack.
(B)   attacking the head of the fire.
(C)   hot spotting.
(D)   cold trailing.

8.   A scratch line is most closely associated with

(A)   hot spotting.
(B)   cold trailing.
(C)   flanking.
(D)   breakover.

9.   If a wildfire is small and fire weather conditions are moderate or lower, the initial attack on the fire should be made at the

(A)   head.
(B)   heel.
(C)   right flank.
(D)   left flank.

10.   In the development of a helispot for use of the copter in a wild-fire, trees and brush should be cleared for a distance of

(A)   100 feet in all directions.
(B)   300 feet in all directions.
(C)   100 feet below the level of the landing area in the approach zone and 300 feet in the departure zone.
(D)   300 feet below the level of the landing area in the approach zone and 100 feet in the departure zone.

Answers

1. (A) The round point shovel is probably the most universally used fire tool. In the hands of a skilled worker, the shovel is very effective. It can be used to dig, to swat, to throw directly, to scrape the fire line clean to mineral soil, and, to some extent, to cut. The shovel also makes an ideal face shield for deflecting heat on a hot fire line. It can be used as a hook to pull material, and it is particularly effective for throwing dirt to check a running fire or to dig out burning material.

2. (D) Conduction is the transfer of heat by direct contact with the heat source. An example is a frying pan on a stove. Wood is ordinarily a poor conductor, but metal is a good conductor. Conduction should be thought of as heat being carried along a conductor as electricity is carried. Although this method of heat transfer is important in structures and flammable liquids, it has little relation to wildfires.

3. (B) Wind is the most variable and least predictable fire weather element affecting wildfires. However, its behavior can be estimated by close study. Winds near the earth's surface are affected by the shape of the topography and by local heating and cooling. There is no substitute for an understanding of local wind behavior.

4. (C) A foehn wind is a special kind of local wind formed by mountain ranges and is a downflowing wind that is warm and dry. These winds flow from a high-pressure area on the windward side of mountains to a low-pressure or trough area on the leeward side. They cause severe fire weather.

    Surface wind speeds of 40 mph to 60 mph are common, and speeds of 90 mph or more have been reported. These winds last for three days or more and usually stop abruptly. The temperature increases rapidly and may change 30° to 40° in a few minutes. The relative humidity often drops quickly to 5% or less with foehn winds.

5. (B) In order of susceptibility, fuels that readily ignite from burning embers are as follows:

    1. Rotten wood, either on the ground, on logs, or in snags.
    2. Moss and lichens in treetops.
    3. Slash, particularly when it is compacted in tight piles.
    4. Duff, peat, dried muck.
    5. Cured grass and grain.

6. (D) Backfiring is a form of indirect attack. It is the process of intentionally starting a fire along the inside edge of a fire line or fire barrier in advance of a head fire or along the forward flanks of a rapidly spreading fire. The area between the control line and the head fire is burned to eliminate fuel in advance of the fire and thus widen the control line, to change the direction of the fire, or to slow the progress of the fire and gain time for line construction. The backfire is usually set considerably in advance of the head fire.

Some area is deliberately sacrificed to prevent a larger fire.

The term backfire is misleading and indicates a simple form of control, which it is not. Fighting fire with fire is hazardous and complex. It should be used only by experienced firefighters and usually as a last resort.

7. (C) Hot spotting, or the point and cutoff technique, is a combination method that may be used in direct attack but is often used in indirect attack. It first attacks the fingers or hot spots along the fire edge by the construction of unconnected lines across those edges that are burning fastest. These short sections of line are then connected, and the flanks are tied into the rear with effective fire lines. More or less, experienced firefighters use this technique in all attacks.

8. (B) Cold trailing is the process of checking out the edge of apparently burned-out sections of the fire. This may mean building or examining a scratch line. A scratch line is a simple line usually about 6 inches wide along a quiet edge of the fire. It is made by beating out or throwing any fire or embers to the inside. Usually, all material taken from the line is thrown inside.

9. (A) If the fire is small and fire weather conditions are moderate or lower, then hit the head with a direct attack. This stops the spread, and the flanks are then worked back to the heel to complete the containment. Attack from the inside if there is little smoke and no danger of burning the tires of the apparatus. Attack from the inside is also best if the wind is turbulent and shifting. However, if the fuel is holding the fire from the initial burn, there may be too much heat and flame to work in the burned area, or conditions may be such that it is unsafe to make a frontal attack. The basic tactic is then to flank or parallel the edge from outside the burn.

10. (C) Trees and brush should be cleared for a distance of 100 feet below the level of the landing area in the approach zone and 300 feet in the departure zone. Additional clearing is needed for larger craft. Landing areas should not be located on the lee side of ridges or any area where a serious fire hazard may exist.

Flammable Liquid and Gas Fires

1. Where fire has involved flow from a liquefied petroleum gas storage tank and the flow cannot be checked, it is generally considered safest practice to

    (A) fight the fire with foam.
    (B) use a dry chemical extinguisher.
    (C) use fog to extinguish the flame and eliminate all
         sources of ignition downwind.
    (D) permit the fire to burn itself out.

2. Ordinary foam will not extinguish fire in alcohol chiefly because

    (A) the alcohol dissolves the foam almost immediately.
    (B) the rate of gas formation is too rapid.
    (C) the foam sinks below the surface.
    (D) the foam is likely to cause a boil-over.

3. When a burning oil tank boils over, it is usually an indication that

    (A) a mixture of oils was present, usually a light oil
         adulterant in heavy oil storage.
    (B) trapped vaporized oil is pushing the boiling oil over
         the tank sides.
    (C) the tank contains light refined oils of low viscosity.
    (D) water is present either in suspended form or at the
         bottom of the tank.

4. "Under normal conditions, do not extinguish a vent fire in an acetylene tank so long as the flame is vigorous and is yellow." This course of action is, in general

    (A) safe; there is little, if any, danger of the flame
         being drawn in through the vent pipe and causing
         an explosion in the tank.
    (B) dangerous; there is great danger of the flame being
         drawn in through the vent pipe and causing an ex-
         plosion in the tank.
    (C) safe; the cylinder is built to withstand explosions
         caused by vent fires.
    (D) dangerous; the cylinder is not built to withstand ex-
         plosions caused by vent fires.

5. "Never extinguish a gas fire, unless by extinguishing it you can gain access to the valve to shut off the flow." This statement is

    (A) correct; in fact it is not possible to extinguish such
         a fire without shutting off the valve.
    (B) not correct; it is important to extinguish the fire as
         soon as possible.
    (C) correct; if gas continues to escape after the flames
         are extinguished there will be the possibility of an
         explosion.
    (D) not correct; the heat produced by the flame is sufficient

to raise the additional fuel above its flash point
and cause an explosion.

6. Of the following flammable gases, which one is lighter than air?

(A) butane
(B) hydrogen sulfide
(C) methyl ether
(D) coal gas

7. Water streams are employed at oil fires to

(A) wet down the oil and thereby extinguish the fire.
(B) cool the atmosphere so that the firefighters can operate at close range.
(C) cover the burning oil surface with a layer of water
and thus smother the fire.
(D) wet down and cool exposed tanks and other oil containers.

8. On a verbal alarm, a pumper was sent to the scene of a truck colli-
sion involving a truck carrying a large number of 55-gallon containers of 200 proof denatured alcohol. There was no fire, but
several containers were leaking, spilling large quantities of alcohol on the street. The officer ordered heavy water streams to
wash the alcohol into the nearest sewer and to wash the truck completely. This action was

(A) good; the large water volume will form a solution that
is not especially hazardous.
(B) poor; if not ignited, the spilled alcohol should have
been covered with foam.
(C) good; the solvent action of alcohol is harmless in a
concrete or brick sewer.
(D) poor; the evaporation of alcohol will prove a serious
explosion hazard in the sewer.

9. In fighting an oil fire which has reached fairly large proportions,
which of the following would you recommend as the best method of
extinguishment?

(A) Direct a forceful stream of water on the fire.
(B) Direct a high velocity stream of foam from a foam
extinguisher into the body of the burning oil.
(C) Direct a stream of foam from a generator at a distance such as to cause it to fall lightly to the
surface of the oil.
(D) Direct a stream of foam from a point as near the
surface of the oil as possible and at such an angle
as to cause it to plunge below the surface of the oil.

10. Containers of compressed gases exposed to a hazardous degree of heat
should preferably be cooled by which of the following?

(A) water
(B) CO_2
(C) carbon tetrachloride
(D) none of the above

Answers

1. (D) Fires involving liquefied petroleum gases should normally not be extinguished unless the flow of the gas can be stopped. If the fire is extinguished without shutting off the flow, the hazard of the continuing escaping gas will be greater than the fire itself. The best procedure is to set up lines to protect the exposures and let the fire burn itself out.

2. (A) Ordinary foam is not effective on fires involving fuels that are water soluble as the liquid immediately dissolves the foam. Some examples of liquids other than alcohol that fall into this category are lacquer thinners, butyl acetate, and acetone. Fighting fires in these liquids requires a special kind of foam known as an "alcohol-type."

3. (D) Boilovers do not occur unless water is present in the oil in either suspended form or on the bottom of the tank. Generally, boilovers occur when the heat wave works itself down until it comes in contact with water at the bottom of the tank, which is quickly converted into steam. The expanding steam forces the oil out of the tank and over the sides, rapidly spreading the fire.

4. (A) It is generally best to allow the acetylene to burn as long as the flame is vigorous and yellow. It is important, however, that it not be allowed to burn too long. Eventually, the refrigerating effect will take place, resulting in a lowering of the internal pressure. This may cause the fire to be drawn into the cylinder. Extinguish the fire before this happens and move the cylinder to a location outside and away from any structure where the gas can be safely dissipated.

5. (C) The problem of extinguishing gas fires without shutting off the flow of the gas falls into the same category as extinguishing a LPG fire without terminating the flow. The same procedure should be used. If the gas cannot be shut off, protect the exposures and let the gas burn until such time that the utility company is able to discontinue the flow.

6. (D) Whether a gas is lighter or heavier than air can be determined from its vapor density. Those gases having a vapor density greater than 1.0 are heavier than air. Those having a vapor density less than 1.0 are lighter than air. The vapor densities of the gases given are: butane--2.0; hydrogen sulfide--1.2; methyl ether--1.6; coal gas--less than 1.0.

7. (D) In some cases water can be used to extinguish oil tank fires; however, in most cases it cannot. Fires in oil tanks, particularly larger tanks, produce a considerable amount of radiated heat. Fire streams should be used to keep adjacent tanks cool. In many situations this requires setting up a number of heavy stream appliances. If hand lines are used, it is generally necessary to set up a device to protect the hose operators from radiant heat while they are playing their lines on exposed tanks.

8. (A) Alcohol is miscible with water, consequently, large quantities of water can be used to dilute it to the point where a hazard no longer exists. Washing the diluted mixture into the storm drain is probably the quickest and most effective method of eliminating the local hazard. The mixture entering the storm drain should not be hazardous.

9. (C) Of the answers given, (C) is undoubtedly the best. The other three alternatives would force the foam or water below the surface of the oil which could increase the boiling effect. Whenever foam is to be applied to the surface of burning oil, it must be delivered in a manner that will permit it to fall or flow gently onto the surface.

10. (A) Water has a higher specific heat than other substances. This means that water has the ability to draw off heat from other bodies faster than other substances. In the protection of exposed containers of compressed gases from radiant heat, water is the best substance to use.

High-Rise Building Fires

Questions 1 through 5 in this section are based upon information contained in the text <u>Fireground Tactics</u> (Emanuel Fried, H. Ginn Corporation, Chicago, 1972). This information has been used with the express permission of the Ginn Corporation.

1. When comparing a bedroom fire in an old flimsy tenement house with fire escapes with one in a modern high-rise building, it can best be said that the

   (A) two fires can be considered exactly alike.
   (B) fire in the high-rise building will provide easier access.
   (C) fire in the high-rise building will be easier to ventilate.
   (D) crew fighting the fire in a high-rise apartment will have to endure a great deal of punishment.

2. There are a number of definitions of high-rise buildings; however, for practical purposes it is best to consider a building a high-rise if

   (A) any portion of the building is above the reach of a department's portable equipment.
   (B) the building is over 100 feet in height.
   (C) the building is over 150 feet in height.
   (D) the building contains more than ten floors.

3. Normally arduous fire duty is further complicated in high-rise fires because of several conditions. Which, if any, of the following is not one of these conditions?

   (A) Line advancement is slow.
   (B) Search and rescue is a time-consuming process.
   (C) Communications between operating units and the chief may be erratic.
   (D) All of the above are conditions making firefighting in a high-rise structure more difficult.

4. The major problem and the one most responsible for the spread of fires in high-rise buildings is the

   (A) kind of interior construction.
   (B) breaking of the windows by the fire allowing the fire to run the exterior.
   (C) lack of horizontal cutoffs.
   (D) flue effect of the vertical openings.

5. One of the lessons learned from the LaSalle and Winecoff high-rise hotel fires is that smoke and heat

   (A) in high-rise buildings will normally be confined to the fire floor.
   (B) will always extend to the top floor when unprotected vertical openings are involved.

(C). will sometimes stratify and consequently not continue to the top of the building, even when unprotected vertical openings are available for fire
travel.

(D) from basement fires will always find its way to the
top floor.

6. The primary concern of the officer--in-charge of a high-rise fire
should be the

(A) location of the fire.
(B) kind of occupancy.
(C) extent of the fire.
(D) life hazard.

7. Which of the following statements regarding the use of elevators
at high-rise fires is the most nearly correct?

(A) Only freight elevators should be used for taking
equipment aloft.
(B) Elevators should not be used until all occupants of
the building have been safely evacuated.
(C) Elevators should not be used until it has been determined that it is safe to do so.
(D) Only one elevator should be used by firefighters. The
others should be reserved for occupant evacuation.

8. One of the most limiting factors in the firefighting efforts at
high-rise fires is the

(A) pressure required to move water to upper floors.
(B) communications problems in relaying orders.
(C) time required to completely evacuate the building.
(D) short period availability of breathing apparatus.

9. It takes about how many firefighters in a high-rise building to
do the job of one firefighter on a similar fire at ground level?

(A) two
(B) three
(C) four
(D) five

10. Both a command post and a staging area should be established in
high-rise firefighting. The most logical location for the establishment of the staging area is

(A) in the street in front of the building.
(B) in the lobby.
(C) two or three floors below the fire.
(D) on the roof.

Answers

1. (D) The good, tight, fire-resistant construction of the high-rise building keeps the heat and smoke inside. If the door is hot to the touch, the firefighters can be sure the inside will be like an oven. If the apartment is large, there may be a long hall (possibly 50 feet long) to traverse before reaching the room or area on fire. The fire may be off the long hall. Firefighters cannot attempt to battle it until they reach the bedroom with the nozzle. Then they have to make a sharp turn to reach in and apply water to the fire. There is no other way to put out such fires. Firefighting crew will have to endure a great deal of punishment.

   It is much easier in an old, flimsy tenement house with fire escapes. Lines can be moved up the fire escape. Ventilation is simple. Just open the windows from the fire escape. There are no exterior fire escapes on modern fire-resistant dwellings; the firefighters must realize that they are in for a rough few minutes.

2. (A) Any fire that is beyond the reach of a department's portable equipment should be considered a high-rise fire. If the department has a 100-foot aerial ladder which can reach up to the eighth or ninth floors, any structure above that height is beyond the department's portable equipment for rescue purposes. If the department uses a ladder pipe from the fly ladder which can penetrate the floor area two stories above the ladder pipe, then for stream penetration anything higher than the eleventh floor is beyond the department's reach and should be considered a high-rise.

3. (D) Line advancement is slow. Firefighters have to bring up rolled lengths of hose by elevators, stopping at the floor below to connect to the standpipe floor valve. They then have to stretch up to the fire floor.

   Search and rescue is a time-consuming process. It takes considerable time for firefighters to search along smoky corridors and rooms for persons who may be trapped. Frequently, rooms are locked, requiring forcible entry with its attendant difficulties and delay.

   Communications between operating units and the chief may be erratic because of the building's steel skeleton construction.

4. (D) The major problem and the one most responsible for the spread of fires in high-rise buildings is the flue effect of the vertical openings such as open interior stairways, elevator shafts, dumbwaiter shafts, and pipe chases. The higher the building, the greater the draft.

5. (C) Smoke and heat in these fires stratified, without reaching the top of the buildings. As a result, loss of life in the LaSalle fire occurred largely below the seventh floor. In like manner, the destructive effect of fire at the Winecoff fire was greatest between the sixth and twelfth floors, and even the toxic effect was less on the fifteenth. It is estimated that the stratification took place because of the cooling of the smoke and heat by the walls as they proceeded upward, causing them to lose their buoyancy at some point.

6.   (D)   Although problems of high-rise fires differ from those at ground level, the primary concern of the officer-in-charge does not change. The saving of life has been, and always will be, the primary objective of firefighting.

7.   (C)   The use of elevators at high-rise fires is a controversial issue. Some departments are questioning whether they should be used at all. Although it is true that the fire could involve elevator shafts, it is unlikely. It would be extremely difficult in some high-rise fires to move firefighters and equipment to upper floors without the use of elevators. However, their use should be restricted until it is determined that they can be used safely.

8.   (B)   Communications has always been one of the major problems at large fires. The problem is even more complicated in high-rise firefighting. Some portable radios are rendered almost useless because of the structure frame of the building. Consequently, it is difficult for the officer-in-charge to relay orders to sector and company commanders, and it is difficult for those fighting the fire to keep the officer-in-charge up-to-date regarding the fire's progress.

9.   (C)   Different departments may establish different figures regarding this relationship; however, four appears to be a good starting place. The difference in required manpower is due primarily to the difficulty of getting to the fire and the additional problem of evacuating occupants.

10.   (C)   Some departments state that the staging area should be established one or two floors below the fire. Although the floor below the fire may be the logical location in some circumstances, this is usually where salvage work is begun. Two or three floors below the fire appears to be a good location for the staging area.

Aircraft Fires

1. The primary objective in an aircraft crash and fire is the

   (A) protection of exposures.
   (B) extinguishment of the fire.
   (C) saving of life.
   (D) keeping the fire loss at an absolute minimum.

2. If a downed aircraft is resting on the side of a hill, if possible, it is best to approach the aircraft from the

   (A) bottom of the hill.
   (B) upper part of the hill.
   (C) windward side of the aircraft.
   (D) leeward side of the aircraft.

3. When advancing fog nozzles for the purpose of making a rescue during crash fire operations, if four nozzlemen are used, the nozzlemen should be positioned so that

   (A) all nozzlemen are on the inside of their hoselines.
   (B) all nozzlemen are on the outside of their hoselines.
   (C) the outside nozzlemen are on the inside of their hoselines and the inside nozzlemen are on the outside of their hoselines.
   (D) the outside nozzlemen are on the outside of their hoselines and the inside nozzlemen are on the inside of their hoselines.

4. When using a protein foam applied with hand lines in crash firefighting, it is best to

   (A) lob the foam onto the fire from maximum range.
   (B) apply the foam in a spray pattern.
   (C) apply the foam to the side of the aircraft.
   (D) direct the foam at the near edge of the fire, let it build up, then push the foam with a sweeping motion until the spill area is covered.

5. Usually, the best method of making entry into an aircraft for the purpose of rescuing trapped people is by

   (A) cutting a hole in the fuselage just behind the pilot's compartment.
   (B) entering through a door in the fuselage.
   (C) cutting a hole in the fuselage about five feet ahead of the tail section
   (D) cutting a hole in the fuselage about midway on the top of the aircraft.

6. Early in a crash fire, it is necessary that

   (A) water be played on all exits of the aircraft.
   (B) water be played on the aircraft's engines.
   (C) water be directed toward the pilot's compartment.

(D)   the entire fuselage be wet down.

7. Where possible, it is usually better to approach an aircraft from the windward side when attempting to gain entrance for rescue purposes. Which of the following explains this approach?

(A)   Greater reach from the fog nozzles is obtained.
(B)   The nozzlemen have better visibility because the smoke from the fire is blowing away from them.
(C)   Greater reach and better visibility is obtained.
(D)   The outlined procedure is not the best approach.

8. If foam and water are used simultaneously on a crash fire, the best method of applying both is to

(A)   use water on the windward side and foam on the leeward side.
(B)   use water on the leeward side and foam on the windward side.
(C)   apply foam from turret nozzles and water by hand lines.
(D)   break a path with water fog nozzles to push the fire and fuel from the aircraft and follow up with foam to secure the path.

9. Where possible, in attempting to gain entrance into the aircraft to effect rescue, the approach should be made

(A)   from the leeside of the aircraft.
(B)   from the windward side of the aircraft.
(C)   toward the nose of the aircraft.
(D)   toward the tail section of the aircraft.

10. It has been found that a group of 1½-inch fog nozzles can be used to provide a good path for rescue purposes at crash fires. When using this kind of attack, it is best to keep the nozzles about

(A)   6 to 12 inches off the ground.
(B)   12 to 18 inches off the ground.
(C)   12 to 24 inches off the ground.
(D)   18 to 36 inches off the ground.

Answers

1. (C) Crash firefighting is no different than other firefighting in its primary objective. The saving of life should always be given first priority. In an aircraft crash fire, this may involve the rescue of people trapped in the aircraft and people in buildings that may become involved in the crash.

2. (B) This is a situation where firefighters should take advantage of gravity to help force the fire from around the aircraft. The best approach is from the top of the hill, using gravity to help drive the fire away from the aircraft. Of course, if there are exposures to the downward side of the hill, it will be necessary to request additional companies to help protect these exposures as burning fuel travels in that direction.

3. (A) It is important when advancing fog lines toward the aircraft for the purpose of making a rescue to provide maximum protection for the nozzlemen. It has been found that maximum protection can be provided by placing all nozzlemen on the inside of their hose-lines.

4. (D) This is the best method of applying protein foam. If the foam is allowed to lob onto the fire, it will break up and be of little use. The foam blanket must be formed in such a manner that the spill or fire will be covered, and remain covered until the emergency has been abated.

5. (B) The best method of gaining entry into the aircraft is through the doors provided. Of course, if the doors cannot be utilized because of the progress of the fire or because they were damaged by the crash, then it will be necessary to cut a hole into the aircraft. Where the hole is cut depends largely upon the situation and the progress of the fire.

6. (D) The entire fuselage should be wet down as soon as possible. It is imperative that the fuselage be kept cool in order to keep the temperature inside the aircraft within survival limits. While cooling down the fuselage, a total effort should be made to drive the flames away from any door that might be used to make entry into the aircraft so that rescue efforts can proceed without delay.

7. (C) The best procedure is to use four nozzlemen to provide a wide pattern of water. The outside nozzlemen should be used to keep the fire away while the inside nozzlemen drive a path toward the opening that will be used for rescue. Approaching from the windward side puts the wind to the nozzlemen's backs and uses it to help push the fire ahead. The wind also helps push the smoke away which provides better visibility for the men making the attack.

8. (D) Using foam and water at the same time on a crash fire is a tricky business. Firefighters must be extremely careful that water streams are not directed in such a manner that a protective foam blanket is broken up. The best procedure is to use a coordinated attack. A four-man fog attack can be made to push the fire and fuel away from the fuselage. This should be followed by nozzlemen

using foam lines to secure the path made by the fog nozzles.

9.   (B)   The approach to the aircraft should always be made from the most advantageous position; however, where possible, the approach should be made from the windward side of the aircraft if there is any amount of wind whatsoever.

10.   (C)   It has been found that the fog nozzle patterns should be interlocked with the nozzles kept 12 to 24 inches off the ground. This attack provides a good fog pattern, and the nozzles are about the right distance above the ground to keep from stirring up any puddles of fuel.

IV

FIRE APPARATUS AND EQUIPMENT

Pumps and Pumping Equipment

Answers to questions in this section are based upon information contained in the text <u>Introduction to Fire Apparatus and Equipment</u> (Gene Mahoney, Allyn and Bacon, Inc., Boston, 1981). This information has been used with the express permission of Allyn and Bacon, Inc.

1. A new positive displacement pump is discharging water at 120 psi. It should be expected that the pump will deliver approximately what percent of its theoretical capacity at this pressure?

    (A) 65% to 75%
    (B) 75% to 85%
    (C) 80% to 90%
    (D) 90% to 95%

2. A compound gage installed on a pumper measures vacuum and pressure respectively in what units?

    (A) feet; pounds
    (B) pounds; feet
    (C) inches; ounces
    (D) inches; pounds

3. The plate on a positive displacement pump reads .79 gpr. The pump is discharging 345 gpm at a pressure of 180 psi while turning at 500 rpm. The volumetric efficiency of this pump under these conditions is most nearly

    (A) 92%.
    (B) 90%.
    (C) 87%.
    (D) 84%.

4. At 200 psi, net pump pressure, a 1,000 gpm pumper will provide approximately

    (A) 450 gpm.
    (B) 525 gpm.
    (C) 575 gpm.
    (D) 700 gpm.

5. Which of the following is not a kind of positive displacement pump?

    (A) rotary gear
    (B) rotary vane
    (C) centrifugal
    (D) piston

6. Pump performance specifications refer to "net pump pressure" at draft, meaning that the discharge pressure is adjusted to compensate for the

    (A) barometric pressure at the time of the test.
    (B) slippage characteristic of the pump.
    (C) lift required to raise water from a static source.
    (D) internal friction loss of the pump.

7. The part of a centrifugal pump that enables the pump to handle an increasing amount of water, but at the same time allows the water to remain at or near the same velocity throughout its entire movement within the pump casing is called the

    (A) gib.
    (B) volute.
    (C) discharge vane.
    (D) impeller shroud.

8. All of the following operations should be performed by the pump operator when charging hose lines, except to

    (A) make certain that hose is fully coupled to the proper pump outlet.
    (B) make certain that the hose crew is ready and waiting for the water.
    (C) open the discharge gate valves quickly.
    (D) open the discharge gate valves slowly.

9. There are three factors involved in a centrifugal pump operation that are so closely related that a change in one will automatically result in a change in another. Which of the following is not one of these three factors?

    (A) pump speed
    (B) discharge pressure
    (C) quantity of discharge
    (D) inlet pressure

10. When operating a centrifugal pump from draft, it is important to open discharge valves slowly. The main purpose of this recommendation is to prevent which of the following?

    (A) development of explosive air pressures in the hose if the shutoff is closed
    (B) cavitation damage to the discharge gate
    (C) water hammer tending to burst the discharge line
    (D) loss of prime

11. One principle involved with centrifugal pumps is that when pumping at a constant quantity, pressure is directly proportional to the square of the pump speed. Using this basic principle, if the pump discharge pressure is 160 psi while the pump is turning at 500 rpm and discharging 300 gpm, what would be the discharge if the pump speed was increased to 650 rpm and the discharge remained the same?

    (A) 200 psi
    (B) 220 psi
    (C) 240 psi
    (D) 270 psi

12. You are operating a pumper from draft with a 20-foot lift and sup-plying a 2½-inch line, 600 feet in length with a 1 1/8-inch tip. You have an abnormally high vacuum reading on the suction gage and are unable to get the proper pressure. This particular condition should suggest to you which of the following as the most likely source of difficulty?

    (A) stoppage at strainer on hard suction
    (B) wrong position of transfer valve
    (C) improper gear of road transmission
    (D) priming pump engaged

13. Each stage of a four-stage centrifugal pump is capable of a capa-city of 150 gpm at 150 psi. If the pump is turning at the speed required to deliver capacity, and the four stages are placed in series, what would be the amount of water and pressure available at the discharge outlets if the incoming pressure is 40 psi?

    (A) 150 gpm at 600 psi
    (B) 150 gpm at 640 psi
    (C) 600 gpm at 150 psi
    (D) 600 gpm at 190 psi

14. The chief difference between a positive displacement pump and a nonpositive displacement pump is that a positive displacement pump

    (A) operates by rotary action while a nonpositive dis-placement pump operates by reciprocating action.
    (B) delivers a definite volume of liquid for each cycle of pump operation, regardless of the resistance offered, while a nonpositive displacement pump de-livers for each cycle, a volume of liquid dependent on the resistance offered.
    (C) operates by a reciprocating action while a nonpositive displacement pump operates by rotary action.
    (D) depends upon cohesion of water molecules for its ac-tion while a nonpositive displacement pump lifts by creating a partial vacuum.

15. Often it is necessary for a pump operator to decide whether to pump in the series or parallel position. Which of the following condi-tions would most likely require the pump operator to pump in the series position?

    (A)   a long single 2½-inch line equipped with a 1¼-inch
          tip
    (B)   three lines taken off the pump each equipped with a
          1-inch tip
    (C)   when supplying a heavy stream appliance
    (D)   whenever pumping more than 50% of rated capacity at
          a net pump pressure of less than 200 psi

16. A centrifugal pumper is connected to a hydrant which is fully open,
     the pump is disengaged, and all discharge gates are closed. The
     pressure reading at the suction gage would be

    (A)   an indication of the number of 2½-inch lines that
          can be supplied.
    (B)   the same as the reading on the discharge gage.
    (C)   less than the reading on the discharge gage.
    (D)   greater than the reading on the discharge gage.

17. Of the following, which is the best indication that a pump is cavi-
     tating?

    (A)   The incoming pressure is reduced to less than 10 psi.
    (B)   The soft intake hose begins to flatten.
    (C)   There is an increase in rpm without an increase in
          pump pressure.
    (D)   The net pump pressure exceeds 250 psi.

18. A pump operator should operate a pump with the changeover or trans-
     fer valve in the position which

    (A)   gives the desired pressure at the least motor speed.
    (B)   gives the desired pressure at the highest motor speed.
    (C)   gives the desired pressure at any motor speed.
    (D)   does not have any bearing on the motor speed.

19. A pump is rated at 1250 gpm at 150 psi, net pump pressure. Which
     is most near the theoretical amount of water this pump can discharge
     at 200 psi net pump pressure?

    (A)   875 gpm
    (B)   937 gpm
    (C)   1,000 gpm
    (D)   1,250 gpm

20. Which of the following statements is correct regarding the opera-
     tion of the transfer valve on a two-stage series parallel centri-
     fugal pump? With the transfer valve in the

    (A)   pressure position, each impeller takes suction and
          discharges water independently.
    (B)   volume or capacity position, the discharge of one
          impeller flows to the suction of the other impeller.
    (C)   pressure position, the impellers operate in series,
          with each impeller developing an equal amount of
          pressure.
    (D)   volume or capacity position, each impeller develops

half of the total discharge pressure.

21. A pump operator is drafting at sea level where the atmospheric pressure is 14.7 psi. The intake gage reads 20 inches Hg. Which is most near the pressure within the main pump?

(A) 5 psi
(B) 7.5 psi
(C) 10 psi
(D) 12 psi

22. What is the name of the pump that depends upon speed to create pressure?

(A) centrifugal
(B) worm gear
(C) rotary gear
(D) piston

23. A pump operator is pumping through two single 2½-inch lines. Once the lines have been provided with the necessary pressure, the pump operator sets the relief valve. If both lines are shut down, it can be expected that the relief valve will hold the pressure within how many psi of the set pressure?

(A) 20
(B) 30
(C) 40
(D) 50

24. When a pumper is being operated, which of the following would best indicate to the operator that the capacity of the hydrant had been reached?

(A) suction gage and discharge gage
(B) transfer valve and discharge gage
(C) discharge gage and tachometer
(D) tachometer and suction gage

25. Most intake and discharge gages on fire apparatus are of the Bourdon type. The principle of operation of this gage is to:

(A) measure the compression of a spring to determine the pressure.
(B) measure the differences of flow between two points to determine the pressure.
(C) use the straightening of a curved tube to measure the pressure.
(D) measure the expansion ratio of a bi-metallic bar to determine the pressure.

26. Which pumps are used on fire apparatus?

(A) centrifugal only
(B) rotary lobe or rotary gear only
(C) piston only

(D)   all of the above

27. A radio-controlled hydrant valve has been attached to a hydrant and a line laid to the fire. From the pumper, the pump operator can

(A)   only open the valve and allow water to flow from the hydrant to the pumper.
(B)   only close the valve when the laid line is no longer needed.
(C)   both open the valve and allow water to flow from the hydrant to the pumper and can close the valve when the water is no longer needed.
(D)   not operate the valve as the valve is designed to shut down automatically in the event of an emergency.

28. How long is the pumping portion of a certification test?

(A)   40 minutes
(B)   2 hours
(C)   3 hours
(D)   4 hours

29. Pumpers are subjected to three different kinds of tests: certification, delivery, and service. The pumping portion of these three tests is the same for which of the following?

(A)   the certification test and delivery test
(B)   the certification test and service test
(C)   the delivery test and service test
(D)   all three of the tests

30. A radio-controlled automatic pumper

(A)   provides for the nozzlemen to control the pressure at the nozzle.
(B)   provides for opening a radio-controlled hydrant valve.
(C)   will automatically open the tank to provide water to the pump in the event the hydrant supply is interrupted.
(D)   does all of the above.

Answers

1. (D) The difference between what a pump discharges and the theoretical amount it should is called slippage. The percentage of slippage of a positive displacement pump depends upon the condition of the pump and the pressure buildup. In general, the higher the pressure, the greater the slippage. Although the slippage of each pump is different, new pumps as a group will deliver 90% to 95% of their theoretical capacity at 120 psi, 80% to 90% at 200 psi, and 75% to 85% at 250 psi.

2. (D) Most pumpers have both an intake gage and a discharge gage. The intake (suction) gage is a compound gage with reading both above and below zero. The readings below zero are in inches of mercury (one inch of mercury is equal to approximately one half psi), while the numbers above zero are in pounds per square inch.

3. (C) Volumetric efficiency is the ratio between the amount of water a pump actually discharges and the theoretical amount it should discharge. Expressed as a formula:

$$\text{Volumetric efficiency} = \frac{\text{discharge X 100}}{\text{theoretical discharge}}$$

The plate on the pump tells the number of gallons per revolution the pump should deliver. The pump given in the problem should deliver .79 gallons per revolution. As the pump is turning at 500 rpm, the theoretical discharge is (500)(.79) = 395 gpm. This information can be inserted into the formula:

$$\text{Volumetric efficiency} = \frac{(345)(100)}{395}$$

$$= \frac{34500}{395}$$

$$= 87.34\%$$

4. (D) Pumps are rated to deliver capacity at 150 psi, net pump pressure; 70% of capacity at 200 psi, net pump pressure; and 50% of capacity at 250 psi, net pump pressure. Therefore, a 1,000 gpm pump would deliver 70% of 1,000 gpm, or 700 gpm at 200 psi, net pump pressure.

5. (C) A positive displacement pump will theoretically discharge a given amount of material with every revolution of the pump shaft; hence, the term positive displacement. These pumps are capable of moving gases as well as liquids which make them ideal for priming pumps. All the pumps listed except the centrifugal pump are kinds of positive displacement pumps.

6. (C) The net pump pressure is the amount of work actually done by the pump. When working from a hydrant, it is the difference between the discharge pressure and the intake pressure. As an example, if the residual pressure on the intake gage is 30 psi while

the discharge pressure is 150 psi, then the net pump pressure is 150 - 30 or 120 psi. The net pump pressure when working from a hydrant is always less than the discharge pressure.

When working from draft the net pump pressure is always greater than the discharge pressure. It takes as much work to lift a given amount of water a given height as it does to push the same amount to the same height. Additionally, the pump must overcome the friction loss in the intake hose when drafting. A pressure correction formula is available for determining the pressure loss due to a given lift. The formula takes into account both the friction loss in the hose and the lift.

7. (B) The volute is a spiral-shaped section of the body of a centrifugal pump that encloses the impeller. It carries water from the impeller to the discharge manifold. The volute is the area between the impeller and the wall of the pump casing. This cross-sectional area constantly increases at a uniform rate as it approaches the discharge outlet.

8. (C) The procedure that is incorrect is the opening of the discharge valve quickly. Discharge valves should be cracked first and water allowed to fill the line. Then, the valve should be opened slowly to the full open position. Opening the line quickly will cause water hammer with the possibility of breaking the line or damaging the pump.

9. (D) The factors are pump speed, discharge pressure, and quantity of discharge. Within limits governed by the pump design, if the pump speed is held constant, an increase in discharge pressure will result in a decrease in the amount of water discharged, and vice versa. If the discharge pressure is held constant, an increase in pump speed will result in an increase in the discharge, and vice versa. If the discharge is held constant, an increase in pump speed will result in an increase in discharge pressure, and vice versa.

10. (D) If a discharge gate is opened too fast when drafting, the pump will try to run away from the water. The result will be the formation of an air pocket within the pump, causing the pump to lose its prime and, consequently, drop the water. It will then be necessary for the apparatus operator to repeat the priming procedure.

11. (D) The principle can be converted to the following formula:

$$\text{new pressure} = \left(\frac{\text{new pump speed}}{\text{old pump speed}}\right)^2 \text{X old pressure}$$

Inserting the information from the problem into the formula:

$$\text{new pressure} = \left(\frac{650}{500}\right)^2 \text{X } 160$$

$$= (1.3)^2 (160)$$
$$= (1.69) (160)$$
$$= 270.4 \text{ psi}$$

12. (A)   There are two primary reasons for having an abnormally high vacuum reading on the suction gage: 1) The lift is too high. 2) The suction strainer is partially obstructed.

13. (B)   A centrifugal pump is capable of taking advantage of incoming pressure. In the problem given, the impeller of stage one would receive the water at 40 psi, add 150 psi of its own, and discharge 150 gpm at 190 psi. The stage two impeller would take the 150 gpm and discharge it into the eye of stage three at 340 psi, adding 150 psi to the 190 psi it received. The stage three impeller would add 150 psi to the 150 gpm and discharge it into the eye of stage four at 490 psi. Stage four would add another 150 psi and discharge the 150 gpm at 640 psi.

14. (B)   A positive displacement pump will theoretically deliver a definite amount of material with every revolution of the pump shaft; a centrifugal pump will not. If a centrifugal pump meets resistance (such as when all discharge gates are closed), it will merely churn, discharging no water.

15. (A)   There are several guidelines which might prove useful when considering whether to pump in the series or parallel position. One general rule is to use the position (series or parallel) that will give the desired result at the lowest engine speed. Another rule is to use the series position when pumping through long single lines and the parallel position when a number of large lines are taken off the pump.

16. (B)   The pathway through a centrifugal pump is continuous. When water is let into the pump with all discharge gates closed, the water will fill the intake manifold, pass through the pump, fill the discharge manifold, and be available at the discharge gates at hydrant pressure. It will also enter both the intake and discharge gages. Until the pump is placed into operation, the reading on both the intake and discharge gage will be the static pressure of the hydrant.

17. (C)   Cavitation occurs whenever the pressure at any point inside a pump drops below the pressure of the water being pumped. Cavitation causes damage which is cumulative and progressive. In reality, cavitation occurs when a pump operator attempts to "run away" from his water. The pump will never cavitate when pumping from a positive water source as long as a positive pressure is maintained on the incoming gage. If the soft intake hose begins to flatten, it is clear that the pump is near the cavitation point. However, the best indication that a pump is cavitating is an increase in rpm without an increase in pump pressure. Whenever this occurs, the throttle should be reduced immediately as any further increase will result in additional damage.

18. (A)   The transfer valve is used on series-parallel centrifugal pumps to switch from series to parallel operation or from parallel operation to series operation. There are several rules in general use for determining whether the transfer valve should be used in the series or parallel position (see Question 15). One of the rules

is to pump in the position that gives the desired result at the lowest engine speed.

19. (B)   The work capability of a pump is expressed in pounds-gallons. The pounds-gallons is determined by multiplying the rated discharge by the rated pressure. In this problem, the working capacity of the pump is equal to (1250)(150) or 187,500 pounds-gallons.

    The theoretical discharge from a pump whenever it is discharging water at a net pump pressure above the rated pressure can be determined by dividing the pounds-gallons capability by the discharge pressure. In this problem, $\frac{187,500}{200} = 937.5$ gpm.

20. (C)   When the transfer valve is in the pressure or series position, the first stage discharges water into the intake of the second stage. Each stage contributes an equal amount to the total discharge pressure. As an example with a two-stage pump, if water is taken into the first stage at zero pressure and discharges 100 gpm at 100 psi into the second stage, the second stage will take the 100 gpm, add its own 100 psi, and discharge 100 gpm at 200 psi.

    When the transfer valve is in the volume or parallel position, each stage takes water independently from the intake source and discharges it independently into the discharge manifold. As an example, if each stage of a two-stage pump is capable of 100 gpm at 100 psi, then each stage would take in 100 gpm and discharge it at 100 psi. The result would be a total discharge of 200 gpm at 100 psi.

21. (A)   Pressure reduction within the main pump is measured in inches of mercury (Hg). One Hg is equivalent to .49 psi, or about ½ psi. When the intake gage reads 20 inches Hg, the pressure reduction within the pump is equal to (20)(.49) or 9.8 psi. Originally, the pressure within the pump was the same as the atmospheric pressure, or 14.7 psi. 14.7 - 9.8 = 4.9. Therefore, the pressure within the pump has been reduced to 4.9 psi.

22. (A)   Positive displacement pumps discharge a given amount of water with each turn of the pump shaft and, in general, turn at a low rpm. On the other hand, centrifugal pumps develop pressure by turning at high speeds. The centrifugal pump is much like swinging a bucket with a hole in the bottom. More water at increased pressure will be discharged through the hole as the bucket is swung faster.

23. (B)   Two kinds of pressure control devices are used in the fire service: relief valves and pressure governors. Relief valves control pressure by opening a bypass between the discharge and suction sides of the pump. Pressure governors control the pressure by controlling the engine throttle setting. Both are designed to operate through a pressure range of approximately 75 psi and 300 psi, and to hold the pressure within 30 psi of the pressure setting if all lines are shut down.

24. (C)   An indication that the capacity of the hydrant has been reached is when there is an increase in rpm (tachometer) without an increase

in pump pressure (discharge gage). At this time, the pump operator is starting to run away from his water. The result is cavitation. Cavitation can cause extreme damage to the pump and should
be avoided. When the operator notices an increase in rpm without
an increase in pump pressure, he should back off a little to stop
the cavitation.

25. (C) The principle of operation of a Bourdon tube is simple. Water
under pressure enters the gage through threaded gage fittings and
passes into a tube. The tube is curved and hollow, and is closed
at the end. The water entering the tube under pressure tends to
straighten it out, resulting in movement at the upper end of the
tube. The degree of movement is used to determine the amount of
pressure.

26. (D) All the kinds of pumps given in the question are used on fire
apparatus; however, today all kinds are not used as main pumps.
All main pumps are now of the centrifugal type. Rotary lobe, rotary gear, and piston pumps are positive displacement pumps. These
pumps are used as booster pumps on some apparatus, and all priming
pumps are of the positive displacement type.

27. (C) Once the valve is attached to the hydrant and the hydrant
opened, the apparatus proceeds to fire with all personnel aboard.
The closed valve holds the water at the hydrant until the pump
operator is ready to use it. When the necessary connections have
been made to the apparatus, the operator pushes a button on the
pump panel marked "hydrant open." A small radio transmitter located on the apparatus sends a signal to the radio-controlled hydrant valve that opens the valve. The pump operator pushes another button marked "hydrant off" whenever he wants to shut down
the hydrant flow.

28. (C) Certification tests are designed to ensure that a new pumper
meets minimum standards of construction and performance prior to
delivery to a purchaser. The pumping portion of the test consists
of drafting water and pumping at rated capacity against a net pump
pressure of 150 psi, followed by two half-hour periods of continuous pumping.

29. (A) The pumping portions for the three tests are as follows:

Delivery and certification tests run for three hours and consist of drafting water and pumping rated capacity against a
net pump pressure of 150 psi for a continuous two-hour period, followed by two half-hour periods of continuous pumping.
During one period, at least 70% of the rated capacity is
delivered at a net pump pressure of 200 psi, and during the
remaining half-hour period, 50% of rated capacity is delivered at a net pump pressure of 250 psi.

The usual service test consists of pumping rated capacity at
150 psi for at least 20 minutes, a pressure test at 70%
capacity at 200 psi for about 10 minutes, and a pressure
test of one-half rated capacity at 260 psi for about 10

minutes.

30. (D) The automatic pumper provides for all of the items listed and then some. The system continuously checks all parameters and makes adjustments as deemed necessary. It monitors such factors as oil pressure, generator, water temperature, governor pressure, water tank level, hydrant pressure, and warning alarms from each nozzleman.

Hydraulics

Answers to questions in this section are based upon information contained in the text Fire Department Hydraulics (Eugene F. Mahoney, Allyn and Bacon, Inc., Boston, 1980). This information has been used with the expressed permission of Allyn and Bacon, Inc.

1. If a pumper at 178 pounds engine pressure is delivering water through 1,325 feet of 2½-inch rubber lined hose equipped with a 1¼-inch nozzle, what, most nearly, is the nozzle pressure in pounds?

    (A) 23
    (B) 29
    (C) 35
    (D) 41

2. The relative carrying capacity of a 30-inch main to a 6-inch main is how many times as much water?

    (A) 10
    (B) 15
    (C) 20
    (D) 25

3. At a fire, it is necessary to flood a basement which measures 10 feet X 10 feet X 12 feet, and estimated to be half full of stock. Two lines of 3-inch hose with open butts are used. With 25 psi at each butt, what, most nearly, is the time it will take to flood the basement completely?

    (A) 1 minute
    (B) 2 minutes
    (C) 3 minutes
    (4) 4 minutes

4. If the "kickback" of a 1¼-inch nozzle delivering water is 120 pounds, what, most nearly, is the nozzle pressure?

    (A) 50 psi
    (B) 58 psi
    (C) 66 psi
    (D) 74 psi

5. One line is stretched from the standpipe on the ninth floor (96 feet above street level) and another from the standpipe on the eighth floor, (84 feet above street level). When solving for nozzle pressure, it is customary to reduce the two lines to a single 2½-inch line with a single nozzle. Doing this, it is necessary to use a single quantity for the back pressure. In the example given, what will be, most nearly, the back pressure in pounds per square inch?

    (A) 19
    (B) 29
    (C) 39
    (D) 49

6. A room 18 feet long and 10 feet wide with a 10-foot ceiling is filled to a depth of 2 feet with fresh water. What is the total water load in pounds on the floor?

    (A) 12,500
    (B) 21,500
    (C) 22,500
    (D) 23,500

7. Of the following statements, which is <u>not</u> an accepted principle of pressures in fluids?

    (A) Fluid pressure at a point in a fluid at rest is of the same intensity in all directions.
    (B) The downward pressure of a liquid in an open vessel varies inversely with the liquid density.
    (C) The downward pressure of a liquid in an open vessel is proportional to its depth.
    (D) The downward pressure of a liquid on the bottom of a vessel is independent of the shape of the vessel.

8. A 500-foot line of 2½-inch rubber-lined hose is stretched from a pumper up a stairway of a building to the fourth floor. A 1 1/8-inch nozzle is used. Allowing 12½ feet per story, what pressure in pounds would be required at the engine to give 40 pounds at the nozzle?

    (A) 110
    (B) 122
    (C) 127
    (D) 148

9. Three nozzles each 1 1/8 inches in diameter are most nearly equivalent to a single nozzle of which diameter?

    (A) 1 1/8 inches
    (B) 2 inches
    (C) 3 inches
    (D) 3 1/3 inches

10. What depth of water, or head, is most nearly equal to that required to produce one pound of pressure per square inch?

    (A) .434 feet
    (B) 1 foot
    (C) 2 feet
    (D) 2.304 feet

11. Suppose that a nozzle on a long line of hose is replaced by a nozzle of smaller diameter, which of the following is the most accurate statement?

    (A) If the stream velocity at the nozzle is reduced, the amount of water discharged is increased.
    (B) A greater flow of water is necessary to prevent the nozzle pressure from falling.

(C) If engine pressure remains the same, nozzle pressure
        is increased.
(D) A longer stretch of line is necessary to prevent the
        nozzle pressure from falling.

12. Which of the following combinations of tips delivering water at the
same nozzle pressure will deliver an amount of water most nearly
equal to the amount delivered by a 1 5/8-inch tip at the same noz-
zle pressure?
(A) five 1/2-inch tips and two 1-inch tips
(B) a 1 1/2-inch tip and a 1 3/4-inch tip
(C) a 1 3/8-inch tip; a 1-inch tip; and a 1/4-inch tip
(D) a 1 1/4-inch tip; a 1 1/8-inch tip; and a 3/4-inch tip

13. The friction loss in three siamesed 3-inch lines, each 1000 feet
long, is most nearly equal to the friction loss in a 2½-inch line
of what length?

(A) 50 feet
(B) 75 feet
(C) 100 feet
(D) 125 feet

14. Using 200 feet of 2½-inch hose, 160-psi pump pressure, and 80 psi
on a 1¼-inch smooth nozzle, what, most nearly, would be the discharge?

(A) 385 gpm
(B) 415 gpm
(C) 445 gpm
(D) 475 gpm

15. A pump at 200 psi is pumping through a single line of 2½-inch hose
1500 feet in length. The nozzle pressure is 48 psi. What is the
K value?

(A) .105
(B) .167
(C) .248
(D) .341

16. The discharge from a 1¼-inch nozzle is 300 gpm. What, most nearly,
is the nozzle pressure?

(A) 32 psi
(B) 42 psi
(C) 52 psi
(D) 62 psi

17. A pumping engine discharges a maximum of 800 gpm at 150 psi. If
specifications require 7/10 rated capacity at 200 psi, what, most
nearly, is the allowance for losses at 200 psi due to friction,
slippage, etc?

(A)   40 gpm
(B)   60 gpm
(C)   100 gpm
(D)   140 gpm

18.   The friction loss of a 1½-inch line is how many times that of a 2½-inch line?

(A)   7
(B)   9
(C)   11
(D)   13

19.   While conducting a test and pumping through two 200-foot lines of 2½-inch hose to a 1 3/8-inch tip at 80 psi nozzle pressure, the soft intake hose ruptured. The intake hose was replaced with a 50-foot section of 3-inch hose. Pumping was resumed to provide the same nozzle pressure at the tip. At this time, what, most nearly, is the friction loss in the 3-inch line between the hydrant outlet and the pumper intake?

(A)   14 psi
(B)   25 psi
(C)   28 psi
(D)   55 psi

20.   Using an angle of deflection of 32º, a nozzle pressure of 45 psi, and a nozzle with a 1 3/8-inch diameter, what, most nearly, will be the maximum effective horizontal reach of a fire stream?

(A)   66 feet
(B)   70 feet
(C)   74 feet
(D)   78 feet

21.   Two engines are each pumping through a single line of 2½-inch hose 1450 feet in length. The two lines are siamesed into a single line of 2½-inch hose 200 feet in length equipped with a 1 3/4-inch nozzle. If the pressure at one engine is 285 psi and 315 psi at the other, what will be the nozzle pressure, most nearly?

(A)   20 psi
(B)   25 psi
(C)   30 psi
(D)   35 psi

22.   Suppose the nozzle pressure on a 2½-inch nozzle is the same as that on a 1-inch nozzle. Then, compared with the discharge from the 1-inch nozzle, what will be the discharge from the 2½-inch nozzle?

(A)   1½ times greater
(B)   2¼ times greater
(C)   3 1/8 times greater
(D)   6¼ times greater

23. If the friction loss in 475 feet of 2½-inch hose is 57 psi, what, most nearly, is the flow?

    (A) 200 gpm
    (B) 225 gpm
    (C) 25 gpm
    (D) 275 gpm

24. A cylindrical tank is 14 feet in diameter and 40 feet in height and is half full of water. How many gallons are there in the tank?

    (A) 18,133
    (B) 19,212
    (C) 23,032
    (D) 38,423

25. "Water being discharged from a nozzle under pressure produces a reaction or 'kick-back'." The amount of this reaction may best be described as varying, for the same nozzle pressure,

    (A) inversely as the square of the nozzle pressure.
    (B) directly as the length of the hose line.
    (C) inversely as the nozzle pressure.
    (D) directly as the square of the nozzle diameter.

26. Suppose a pumper with an engine pressure of 200 psi delivers water through two parallel lines of 2½-inch hose each 400 feet long, siamesed into a deluge set equipped with a 1½-inch nozzle. What, most nearly, would be the nozzle pressure?

    (A) 84 psi
    (B) 92 psi
    (C) 100 psi
    (D) 108 psi

27. An engine is pumping through a 500-foot line of 2½-inch hose which is laid into the siamese connection of a standpipe. A 100-foot line of 2½-inch hose is stretched from the outlet on the twelfth floor and is equipped with a 1-inch nozzle. Allowing 12 feet per story, what, most nearly, would be the engine pressure required to provide a nozzle pressure of 50 psi?

    (A) 143 psi
    (B) 175 psi
    (C) 200 psi
    (D) 225 psi

28. What, most nearly, is the friction loss in 200 feet of 3-inch hose when the flow is 250 gpm?

    (A) 8 psi
    (B) 10 psi
    (C) 12 psi
    (D) 14 psi

29. An engine at 250 psi is pumping through a 100-foot line of 2½-inch hose that is breached to three parallel lines of 2½-inch hose, each 300 feet long. Each of the three lines is equipped with a 1 1/8-inch nozzle. What, most nearly, is the nozzle pressure at each nozzle?

   (A)   36 psi
   (B)   43 psi
   (C)   46 psi
   (D)   58 psi

30. If four siamesed lines of 2½-inch hose, two of which are 1,000 feet each, the others 1,200 feet each, are together carrying 850 gpm, what, most nearly, would be the total friction loss in the layout?

   (A)   71 psi
   (B)   108 psi
   (C)   128 psi
   (D)   135 psi

Answers

1. (A)

$$NP = \frac{EP}{1.1 + KL}$$

Where

$$EP = 178$$
$$K = .248$$
$$L = 26.5$$

Then

$$NP = \frac{178}{1.1 + (.248)(26.5)}$$

$$= \frac{178}{1.1 + 6.57}$$

$$= \frac{178}{7.67}$$

$$= 23.2 \text{ psi}$$

2. (D)   A problem like this can usually be solved by comparing the square of the diameters.

$$\frac{D^2}{d^2}$$

Where

$$D = 30 \text{ inches}$$
$$d = 6 \text{ inches}$$

Then

$$\frac{(30)^2}{(6)^2} = \frac{900}{36}$$

$$= 25$$

3. (B)   First find the number of gallons of water required to flood the basement by determining the volume of the basement and multiplying it by the number of gallons in a cubic foot (7.48). Note: As the basement is half filled with stock, the space to be flooded measures 10 feet by 10 feet by 6 feet.

$$\text{Gallons} = 7.48 \text{ V}$$

Where

$$V = LWH$$

And

$$L = 10 \text{ feet}$$
$$W = 10 \text{ feet}$$
$$H = 6 \text{ feet}$$

Then

$$V = (10)(10)(6)$$
$$= 600 \text{ cubic feet}$$
$$\text{Gallons} = 7.48 \text{ V}$$
$$= (7.48)(600)$$
$$= 4,488$$

Next find the discharge from the two 3-inch open butt lines

$$\text{Discharge (from one)} = 27\ D^2\sqrt{P}$$

Where

$$D = 3 \text{ inches}$$
$$P = 25$$

Then

$$\text{Dis} = (27)(3)(3)\ \sqrt{25}$$
$$= (27)(9)(5)$$
$$= 1{,}215 \text{ gpm}$$

Then two lines

$$= (2)\ (1215)$$
$$= 2430$$

Now divide.

$$\text{Time} = \frac{4488}{2430}$$
$$= 1.85 \text{ minutes}$$

4. (A) The formula for nozzle reaction is

$$NR = 1.5\ D^2 P$$

This formula can be manipulated to establish a formula for P:

$$NR = 1.5\ D^2 P$$
$$1.5\ D^2 P = NR$$
$$P = \frac{NR}{1.5\ D^2}$$

Where

$$NR = 120$$
$$D = 1.25$$

Then

$$P = \frac{120}{(1.5)(1.25)(1.25)}$$
$$= \frac{120}{2.34}$$
$$= 51.3 \text{ psi}$$

5. (C) When working this kind of problem, the average back pressure should be used.

$$96 \text{ feet} + 84 \text{ feet} = 180 \text{ feet}$$
$$180 \text{ feet} \div 2 = 90 \text{ feet}$$

The formula for back pressure is:

$$BP = .434\ H$$

Where

$$H = 90$$

Then

$$BP = (.434)(90)$$
$$= 39.06 \text{ psi}$$

6. (C) This problem can be solved by finding the volume of water in the room and multiplying it by the weight per cubic foot (62.5).

$$V = LWH$$

Where

$$L = 18 \text{ feet}$$
$$W = 10 \text{ feet}$$
$$H = 2 \text{ feet}$$

Then

$$V = (18)(10)(2)$$
$$= 360 \text{ cubic feet}$$

Then

$$\text{weight} = (62.5)(360)$$
$$= 22,500 \text{ pounds}$$

7. (B) The six basic rules governing the pressure in liquids are as follows:

1. Fluid pressure is perpendicular to any surface on which it acts.
2. The pressure at any point in a fluid at rest is of the same intensity in all directions.
3. External pressure applied to a confined liquid is transmitted undiminished in all directions.
4. The downward pressure of a liquid in an open container is directly proportional to the depth of the liquid.
5. The downward pressure of a liquid in an open container is directly proportional to its density.
6. The downward pressure of a liquid on the bottom of an open container is independent of the shape or size of the container.

    Answer (A) is rule 2.
    Answer (B) is not a rule.
    Answer (C) is rule 5.
    Answer (D) is rule 6.

8. (C)

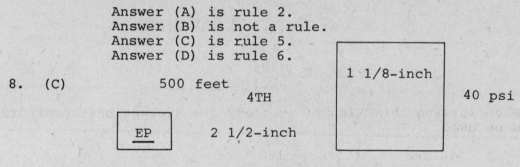

$$REP = EP + BP$$

Take the layout to ground level:

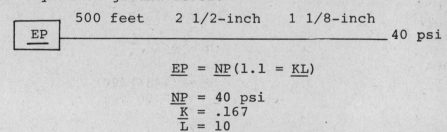

$$EP = NP(1.1 = KL)$$

Where

$$NP = 40 \text{ psi}$$
$$K = .167$$
$$L = 10$$

Then

$$EP = 40 \left[1.1 + (.167)(10) \right]$$
$$= 40(1.1 + 1.67)$$
$$= (40)(2.77)$$
$$= 110.8 \text{ psi}$$

Next solve for the back pressure. The line is being worked three floors above street level:

Where

$$BP = .434 H$$

$$H = (3)(12.5)$$
$$= 37.5 \text{ feet}$$

Then

$$BP = (.434)(37.5)$$
$$= 16.28 \text{ psi}$$

Then

$$REP = EP + BP$$
$$= 110.8 + 16.28$$
$$= 127.08 \text{ psi}$$

9. (B) This kind of problem can be solved by the 8's method.

First, change tips to 8's:

$$1 \ 1/8 = \frac{9}{8}$$

Then square the 8's:

$$9^2 = \frac{81}{8}$$

Then add all the tips together:

$$\frac{81}{8} + \frac{81}{8} + \frac{81}{8} = \frac{243}{8}$$

Now take the square roots of the 8's

$$\frac{\sqrt{243}}{8} = \frac{15.59}{8}$$

The nearest tip size would be

$$\frac{16}{8} \text{ or 2 inches}$$

10. (D) The formula to use to solve the problem is:

$$H = 2.304 \ P$$

Where

$$P = 1$$

Then

$$H = (2.304)(1)$$
$$= 2.304 \text{ feet}$$

11. (C) Compare the following layouts where the engine pressure remains the same:

$$\boxed{EP} \underset{\text{400 feet}}{\overline{\hspace{3cm}}} \overset{\text{2 1/2-inch}}{\hspace{3cm}} \text{1 1/4-inch}$$

$$NP = \frac{EP}{1.1 + KL}$$

$$= \frac{200}{1.1 + (.248)(8)}$$

$$= \frac{200}{1.1 + 1.98}$$

$$= \frac{200}{3.08}$$

$$= 64.94 \text{ psi}$$

$$\boxed{EP} \underset{\text{400 feet}}{\overline{\hspace{3cm}}} \overset{\text{2 1/2-inch}}{\hspace{3cm}} \text{1-inch}$$

$$NP = \frac{EP}{1.1 + KL}$$

$$= \frac{200}{1.1 + (.105)(8)}$$

$$= \frac{200}{1.1 + 0.84}$$

$$= \frac{200}{1.94}$$

$$= 1.03.1 \text{ psi}$$

This sample shows that answer (C) is correct.

12. (C) Determine the equivalent nozzle diameter of each of the combinations by the 8's method (see question 9).

(A) Five ½-inch $= \frac{80}{8}$; two 1-inch $= \frac{128}{8}$; then $\frac{80}{8} + \frac{128}{8} = \frac{208}{8}$

$\frac{\sqrt{208}}{8} = \frac{14.42}{8}$

(B) A 1½-inch $= \frac{144}{8}$; a 1¼-inch $= \frac{100}{8}$; then $\frac{144}{8} + \frac{100}{8} = \frac{244}{8}$

$\frac{\sqrt{244}}{8} = \frac{15.6}{8}$

(C) A 1 3/8-inch $= \frac{121}{8}$; 1-inch $= \frac{64}{8}$; ¼-inch $= \frac{4}{8}$; then $\frac{121}{8} +$

$\frac{64}{8}$ $\frac{4}{8}$ $\frac{189}{8}$ $\frac{\sqrt{189}}{8} = \frac{13.74}{8}$

(D)   A 1¼-inch = $\frac{100}{8}$; 1 1/8-inch = $\frac{81}{8}$; 3/4-inch = $\frac{36}{8}$; then

$$\frac{100}{8} + \frac{81}{8} + \frac{36}{8} = \frac{217}{8} \quad \frac{\sqrt{217}}{8} = \frac{14.73}{8}$$

A 1 5/8-inch tip = $\frac{13}{8}$

The combination of tips in answer (C) is closest to 13.

13.   (A)   The friction factor for three siamesed 3-inch lines is 20.4. To determine the equivalent length of 2½-inch hose, divide the average length of the siamese by the factor:

$$\frac{1000}{20.4} = 49 \text{ feet}$$

14.   (B)   This problem has a lot of unnecessary information.

Simply use the discharge formula:

$$\text{Dis.} = 30 \, D^2 \sqrt{P}$$

Where

$$D = 1.25$$
$$P = 80$$

Then

$$\text{Dis.} = (30)(1.25)^2 \sqrt{80}$$
$$= (30)(1.56)(8.94)$$
$$= 418.39 \text{ gpm}$$

15.   (A)   The engine pressure formula can be manipulated to determine a formula for K:

$$EP = NP(1.1 + KL)$$

$$NP(1.1 + KL) = EP$$

$$1.1 + KL = \frac{EP}{NP}$$

$$KL = \frac{EP}{NP} - 1.1$$

$$K = \frac{\frac{EP}{NP} - 1.1}{L}$$

In this problem:

$$EP = 200$$
$$NP = 48$$
$$L = 30$$

Then

$$K = \frac{\frac{200}{48} - 1.1}{30}$$

$$= \frac{4.17 - 1.1}{30}$$

$$= \frac{3.07}{30}$$

$$= .102$$

16. (B) The discharge formula can be manipulated to create a formula for finding the nozzle pressure:

$$Dis. = 30D^2 \sqrt{P}$$

$$30D^2 \sqrt{P} = Dis.$$

$$\sqrt{P} = \frac{Dis.}{30D^2}$$

$$P = \left(\frac{Dis.}{30D^2}\right)^2$$

Where

$$Dis. = 300 \text{ gpm}$$

$$D = 1.25$$

Then

$$P = \left(\frac{300}{(30)(1.25)(1.25)}\right)^2$$

$$= \left(\frac{300}{46.88}\right)^2$$

$$= (6.4)^2$$

$$= 40.96$$

17. (A) The theoretical work capacity of the pump can be determined by multiplying rated capacity by rated pressure.

$$(800)(150) = 120,000 \text{ pounds per gallons}$$

The theoretical discharge at a pressure above the rated pressure can be determined by dividing the pounds per gallons capability by the new pressure:

$$\frac{120,000}{200} = 600$$

The specifications say 70% at 200 psi or

$$(.70)(800) = 560 \text{ gpm}$$

The difference between the theoretical discharge and the specification discharge (40 gpm) is the result of friction, slippage, etc.

18. (D) This problem can be solved by referring to the friction loss supplement for a single 1½-inch line, which is 13.

19. (A) First, find the discharge from a 1 3/8-inch tip at 80 psi:

$$Dis. = 30\underline{D}^2 \sqrt{\underline{P}}$$

Where

$$\underline{D} = 1 \ 3/8 \ or \ 11/8$$
$$\underline{P} = 80$$

Then

$$Dis. = (30) \left(\frac{11}{8}\right) \left(\frac{11}{8}\right) \sqrt{80}$$

$$= \frac{(30)(11)(11)(8.94)}{64}$$

$$= \frac{32452.2}{64}$$

$$= 507 \ gpm$$

Then find the friction loss in 100 feet of 2½-inch hose when the flow is 507 gpm:

$$\underline{FL} = 2 \ Q^2 + Q$$

$$= (2)(5.07)(5.07) = 5.07$$

$$= 51.41 + 5.07$$

$$= 56.48 \ psi$$

The friction loss in 50 feet of 2½-inch hose = $\frac{56.48}{2}$ or 28.24.

The friction factor for a single 3-inch line is 2.6. Therefore, the friction loss in 50 feet of 3-inch hose equals:

$$\frac{28.24}{2.6} = 10.86 \ psi$$

20. (A) The formula for finding horizontal reach is:

$$\underline{S} = \sqrt{(\underline{HF})(\underline{P})}$$
$$\underline{HF} = horizontal \ factor$$
$$\underline{P} = nozzle \ pressure$$

The horizontal factor for a 1 3/8-inch tip is 98.

Then

$$\underline{S} = \sqrt{(98)(45)}$$

$$= \sqrt{4410}$$

$$= 66.4 \ feet$$

21.   (B)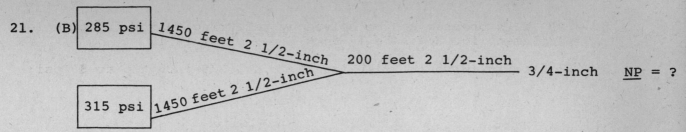

First, average the pump pressures and consider them as one pumper:

$$285 + 315 = 600$$

$$\frac{600}{2} = 300$$

The layout then appears as follows:

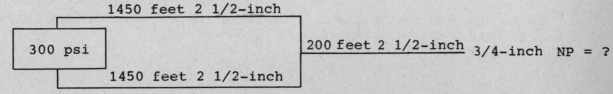

Next, change the layout to a single 2½-inch equivalent:

The friction factor for two 2½-inch siamesed lines is 3.6.

$$\frac{1450}{3.6} + 200 = 403 + 200 = 603 \text{ feet}$$

The layout then appears as follows:

```
 ┌──────────┐  603 feet 2 1/2-inch
 │ 300 psi  │──────────────────────── 3/4-inch    NP = ?
 └──────────┘
```

Then

$$NP = \frac{EP}{1.1 + KL}$$

Where

EP = 300 psi
K = .907
L = 12.06

Then

$$NP = \frac{300}{1.1 + (.907)(12.06)}$$

$$= \frac{300}{1.1 + 10.94}$$

$$= \frac{300}{12.04}$$

$$= 24.92 \text{ psi}$$

22.   (D) When the pressure on two tips is the same, the relationship between the flows can be determined by comparing the square of the tip diameters.

$$\frac{D^2}{d^2}$$

In this problem

$$D = 2.5$$
$$d = 1$$

Then

$$\frac{(2.5)^2}{(1)^2} = \frac{6.25}{1} \text{ or } 6.25$$

23. (B) The formula for determining the flow when the friction loss is known is:

$$Q = \frac{\sqrt{2\ FL}}{2} - 0.25$$

FL in the formula is the friction loss in 100 feet of single 2½-inch hose. In this problem, FL equals:

$$\frac{57}{4.75} = 12 \text{ psi}$$

To solve:

$$Q = \frac{\sqrt{(2)(12)}}{2} - 0.25$$

$$= \frac{\sqrt{24}}{2} - 0.25$$

$$= \frac{4.90}{2} - 0.25$$

$$= 2.45 - 0.25$$

$$= 2.25 \text{ or } 225 \text{ gpm}$$

24. (C) First, determine the volume of the cylindrical container (water space only):

$$V = .7854\ D^2H$$

Where

$$D = 14$$
$$H = 20$$

Then

$$V = (.7854)(14)(14)(20)$$
$$= 3078.77 \text{ cubic feet}$$

Next, multiply the gallons per cubic foot by the volume:

$$\text{gallons} = 7.48\ V$$
$$= (7.48)(3078.77)$$
$$= 23,029.2$$

25. (D) The formula for nozzle reaction is: $NR = 1.5\ D^2P$. This means that the nozzle reaction is 1.5P times the square of the nozzle diameter, or that it will vary directly (nozzle reaction increases as the diameter increases) as the square of the nozzle diameter.

26.   (B)

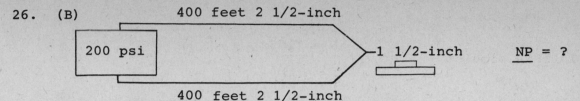

First, change the layout to an equivalent lay of single 2½-inch
hose. The friction factor for 2-2½-inch lines is 3.6.

$$\frac{400}{3.6} = 111 \text{ feet}$$

The layout then appears as follows:

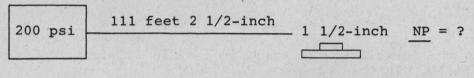

Then

$$NP = \frac{EP}{1.1 + KL}$$

Where

$$EP = 200$$
$$K = .505$$
$$L = 2.22$$

Then

$$NP = \frac{200}{1.1 + (.505)(2.22)}$$

$$= \frac{200}{1.1 + 1.12}$$

$$= \frac{200}{2.22}$$

$$= 90.1 \text{ psi}$$

27.   (C)

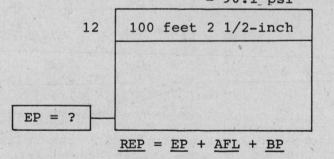

$$REP = EP + AFL + BP$$

First, remove the layout to the ground level:

EP = ? ———— 600 feet 2 1/2-inch ———— 1-inch   50 psi

Then

$$EP = NP(1.1 + KL)$$

Where

$$NP = 50$$
$$K = .105$$
$$L = 12$$

Then

$$EP = 50 \quad 1.1 + (.105)(12)$$
$$= 50(1.1 + 1.26)$$
$$= (50)(2.36)$$
$$= 118 \text{ psi}$$

Then, find the head and back pressure:

$$H = (11)(12)$$
$$= 132 \text{ feet}$$
$$BP = .434 \text{ H}$$
$$= (.434)(132)$$
$$= 57.29 \text{ psi}$$

The appliance friction loss for a standpipe is 25 psi.

Then

$$REP = EP + AFL + BP$$
$$= 118 + 25 + 57.29$$
$$= 200.29 \text{ psi}$$

28. (C) First, determine the friction loss in 200 feet of single 2½-inch hose when the flow is 250 gpm:

$$FL = 2 \ Q^2 + Q$$
$$= (2)(2.5)(2.5) = 2.5$$
$$= (12.5 + 2.5$$
$$= 15 \text{ psi}$$

The friction loss in 200 feet of 2½-inch hose = (2)(15) = 30 psi.

Next, divide the friction loss in the 2½-inch line by the friction factor for a single 3-inch line (2.6):

$$\frac{30}{2.6} = 11.54 \text{ psi}$$

29. (C)

Note: As all lines are of the same length, the nozzle pressure on each line will be the same.

First, find the equivalent nozzle diameter:

$$\frac{81}{8} + \frac{81}{8} + \frac{81}{8} = \frac{243}{8} \qquad \frac{\sqrt{243}}{8} = \frac{15.59}{8} \text{ or nearest 2 inches}$$

Next, change the three 2½-inch siamesed lines to a single 2½-inch equivalent. The friction factor for three 2½=inch line is 7.75:

$$\frac{300}{7.75} = 39 \text{ feet}$$

The total layout equivalent is: 100 + 39 = 139 feet

The layout now appears as follows:

$$NP = \frac{EP}{1.1 + KL}$$

Where

$$EP = 250 \text{ psi}$$
$$K = 1.55$$
$$L = 2.78$$

Then

$$NP = \frac{250}{1.1 + (1.55)(2.78)}$$

$$= \frac{250}{1.1 + 4.31}$$

$$= \frac{250}{5.41}$$

$$= 46.2 \text{ psi}$$

30. (D) First, determine the average length of the siamese:

$$\frac{1000 + 1000 + 1200 + 1200}{4} = \frac{4400}{4} = 1100 \text{ feet}$$

Next, determine the friction loss in 100 feet of 2½-inch hose carrying 850 gpm:

$$FL = 2 Q^2 + Q$$
$$= (2)(8.5)(8.5) + 8.5$$
$$= 144.5 + 8.5$$
$$= 153 \text{ psi}$$

Then, determine the friction loss in 1100 feet of 2½-inch hose:

$$(11)(153) = 1683 \text{ psi}$$

Next, divide the friction loss in the four 2½-inch lines by the friction factor for four 2½-inch lines (12.4):

$$\frac{1683}{12.4} = 135.73 \text{ psi}$$

Fire Apparatus

Answers to questions in this section are based upon information contained in the text <u>Introduction to Fire Apparatus and Equipment</u> (Gene Mahoney, Allyn and Bacon, Inc., Boston, 1981). This information has been used with the express permission of Allyn and Bacon, Inc.

1. During an oil change, the amount of oil required to refill the apparatus engine crankcase is substantially less than originally required. This indicates that

   (A) the engine is not burning oil.
   (B) gasoline is leaking into the crankcase.
   (C) the oil filter should be changed.
   (D) sludge is present in the oil pan.

2. A pumping engine is capable of raising 2,000 gallons per minute a vertical distance of 100 feet. Neglecting friction and other losses, what, most nearly, is the horsepower that the engine delivers during this operation?

   (A) 40
   (B) 50
   (C) 60
   (D) 70

3. If the first stroke of a four-stroke-cycle engine is the power stroke, what would be the fourth stroke?

   (A) intake stroke
   (B) exhaust stroke
   (C) power stroke
   (D) compression stroke

4. Detonation and pre-ignition are two kinds of abnormal combustion. These two phenomena are quite

   (A) different in origin, but very similar in effect.
   (B) different in origin and very different in effect.
   (C) similar in origin and very similar in effect.
   (D) similar in origin, but very different in effect.

5. The primary purpose of the flywheel in an engine is to

   (A) assist in providing smooth engine operation.
   (B) serve as a base for the clutch.
   (C) turn the crankshaft when the engine is started.
   (D) turn auxiliary equipment.

6. The compression ratios of diesel engines used on fire apparatus as compared with gasoline engines are

   (A) slightly less than those in gasoline engines.
   (B) about the same as those in gasoline engines.
   (C) about twice those in gasoline engines.
   (D) about three times those in gasoline engines.

7. What is the engine displacement of an eight-cylinder engine which has a 5-inch bore and a 6-inch stroke?

   (A)   753 cubic inches
   (B)   824 cubic inches
   (C)   886 cubic inches
   (D)   952 cubic inches

8. If the air-fuel ratio provided by the carburetor is 14 to 1, this means that there are 14

   (A)   cubic feet of air burned to 1 cubic foot of gasoline.
   (B)   gallons of air burned to 1 gallon of gasoline.
   (C)   pounds of air burned to 1 pound of gasoline.
   (D)   pounds of air burned to 1 gallon of gasoline.

9.

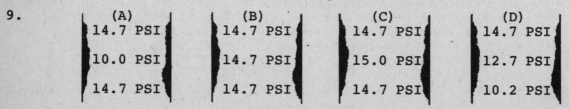

(A)	(B)	(C)	(D)
14.7 PSI	14.7 PSI	14.7 PSI	14.7 PSI
10.0 PSI	14.7 PSI	15.0 PSI	12.7 PSI
14.7 PSI	14.7 PSI	14.7 PSI	10.2 PSI

   Above are shown four possible conditions that might occur when a venturi is present in a path of flow. Which most nearly indicates the principle of a venturi?

10. When is the power enrichment system within a carburetor used?

   (A)   when the engine is idling
   (B)   when the engine is first started
   (C)   for rapid accelerating
   (D)   at high speeds

11. A clutch brake on a fire apparatus

   (A)   is used when shifting up.
   (B)   is used when shifting down.
   (C)   is used when shifting both up and down.
   (D)   has nothing to do with upshifts or downshifts.

12. What is the primary advantage of disc brakes?

   (A)   increased braking surface
   (B)   reduced brake fading
   (C)   less cost
   (D)   lower maintenance

13. Which of the following is not one of the units found in a typical regulator assembly?

   (A)   reverse current condensor
   (B)   cut-out relay
   (C)   current regulator
   (D)   voltage regulator

14. How is the length of an aerial ladder measured?

   (A) from the turntable to the top rung when the ladder
       is fully extended
   (B) from the turntable to the top rung plus the dis-
       tance from the turntable to the ground when the
       ladder is fully extended
   (C) by a plumb line which extends from the top rung to
       the ground when the ladder is fully extended at
       its maximum elevation
   (D) from the top rung to the ground along the plane of
       the ladder when the ladder is fully extended

15. Which kind of aerial ladder apparatus is the most maneuverable?

   (A) rear-mounted aerial
   (B) midship-mounted aerial
   (C) tractor-trailer combination
   (D) All of the above are approximately equivalent in
       maneuverability.

16. Aerial ladders are generally designed using a truss construction.
    This construction provides

   (A) greater strength when operated in a tension situa-
       tion.
   (B) greater strength when operated in a compression sit-
       uation.
   (C) almost equal strength in either a tension or compres-
       sion situation.

17. Levers used to operate the various functions of an aerial ladder
    are generally self-centering. This most nearly means that they

   (A) return to neutral in the event the operator's hand
       slips off the control.
   (B) must be returned to neutral at the end of an opera-
       tion before the next operation is commenced.
   (C) are controlled by a lock which prohibits their move-
       ment from the neutral position until the control is
       unlocked.
   (D) must be forced out of and into the neutral position.

18. When operating an aerial ladder, the most stress will be placed on
    the apparatus when the ladder is

   (A) fully extended, at maximum elevation, in an unsupported
       position.
   (B) fully extended, at maximum elevation in a supported
       position.
   (C) fully extended in a cantilever position horizontal to
       the plane of the earth.

      (D)   raised in-line with the apparatus in a fully extended
           position at approximately a 45° angle of elevation.

19.   The inclinometer on an aerial ladder is used primarily to inform
     the apparatus operator of the

     (A)   operating limitations.
     (B)   angle of operation.
     (C)   ladder's extension.
     (D)   load placed on the ladder.

20.   Jackknifing is used to stabilize a tractor-trailer aerial ladder
     apparatus. The ideal jackknife is a(n)

     (A)   outside jack which provides a 45° angle between the
           tractor and the trailer.
     (B)   inside jack which provides a 45° angle between the
           tractor and the trailer.
     (C)   outside jack which provides a 60° angle between the
           tractor and the trailer.
     (D)   inside jack which provides a 60° angle between the
           tractor and the trailer.

21.   Which is the longest elevated platform in use in the United States?

     (A)   articulated type
     (B)   telescopic type
     (C)   combination telescopic-articulated type
     (D)   The maximum length of each of the above types is ap-
           proximately the same.

22.   The effective working height of an elevated platform is measured
     from the

     (A)   turntable, along the booms to the floor of the plat-
           form.
     (B)   ground, along the booms to the platform.
     (C)   floor of the platform to the ground, by a plumb line.
     (D)   upper railing of the platform to the ground, by a
           plumb line.

23.   When comparing the horizontal reach of an 85-foot aerial ladder
     with an 85-foot articulated elevated platform, it can best be said
     that the horizontal reach of

     (A)   each is about the same.
     (B)   the elevated platform is slightly greater.
     (C)   the aerial ladder is slightly greater.
     (D)   the aerial ladder is approximately twice that of
           the elevated platform.

24.   It can best be said that the platform-carrying capacity of elevat-
     ed platforms varies between which of the following?

     (A)   300 and 700 pounds
     (B)   500 and 800 pounds

(C)   700 and 1000 pounds
(D)   1200 and 1500 pounds

25.   Deadman switches on aerial platform apparatus are most closely
      associated with which of the following?

      (A)   stabilizing devices
      (B)   booms
      (C)   basket levelling system
      (D)   turntable

Answers

1. (D) The oil pan is the reservoir for the engine oil. The dipstick, used to determine the amount of oil in the engine, measures the amount of oil in the pan. When the bottom of the pan becomes loaded with sludge, the amount of oil in the pan is actually less than that indicated on the dip stick. Consequently, when the oil is changed, the amount required to bring it up to the full mark on the dipstick will be less than if there was no sludge in the bottom of the pan.

2. (B) Horsepower is the force required to raise 33,000 pounds at the rate of one foot in one minute. The horsepower required to perform a given amount of work can be determined by the formula:

$$horsepower = \frac{DW}{33,000T}$$

Where

D = distance through which W must be moved
W = the weight that is to be moved through D
T = the time required to move W through D (in minutes)

In this problem

D = 100 feet
W = 2000 X 8.35 = 16700 (water weighs 8.35 pounds per gallon)
T = one minute

Then

$$hp = \frac{(100)(16700)}{(33000)(1)}$$

$$= \frac{1670000}{33000}$$

$$= 50.6$$

3. (D) The four strokes of a four-stroke-cycle engine given in the proper sequence are: intake, compression, power, exhaust. If these are arranged with the power stroke as stroke number one, they appear as follows: power, exhaust, intake, compression.

4. (A) An example of abnormal combustion occurs when two flame fronts collide in the combustion chamber. This can be caused by either detonation or pre-ignition, two phenomena that are quite different in origin, but very similar in effect. One can cause the other, and either is capable of causing severe damage within the combustion chamber.

Detonation is the ignition of a second flame after the timed spark has occurred. It is caused by the use of the wrong octane fuel, a portion of which starts to burn spontaneously from increased heat and pressure. It can also be caused by factors such as excessive temperature, advanced timing, engine lugging, increased compression rating from deposit buildup, and induction leaks.

Pre-ignition is the ignition of a fuel charge before the spark ignites. It is caused by hot spots within the cylinder, such as overheated plugs, glowing depositions, improperly seated valves, or rough metal edges in the chamber.

5. (A) The primary purpose of the flywheel is to assist in providing smooth engine operation. Power to the crankshaft varies from second to second according to the timing of the power strokes of the various cylinders. Although a power overlap occurs to some extent between the different cylinders, there are still variations in the amount of power being transmitted to the crankshaft. The crankshaft tries to speed up when the power is the greater and slow down when it is less. The flywheel tends to resist any change in speed by storing power when the power is greater and releasing it through its inertia during periods of less power.

6. (D) The compression ratios of diesel engines are much higher than those of gasoline engines. Although the gasoline engines used in fire apparatus have compression ratios of about 8 to 1, those in diesel engines approach 21 to 1. These high compression ratios provide pressures of about 600 psi at the end of the compression stroke, raising the temperature of the compressed air to approximately 1000°F. This temperature is sufficient to ignite the diesel fuel without any additional ignition source, thereby eliminating completely the ignition system used in gasoline engines.

7. (D) Engine displacement is a measurement of the size of an engine. It is the sum total of the piston displacement of all cylinders within the engine. Engine displacement can be found by using the following formula:

$$\text{engine displacement} = ASN$$

Where
A = area of a piston (square inches)
S = stroke (inches)
N = number of cylinders

In this problem

$$A = .7854 \, D^2$$
$$= (.7854)(5)(5)$$
$$= 19.635$$

Then

$$\text{engine displacement} = (19.635)(6)(8)$$
$$= 942.48 \text{ cubic inches}$$

8. (C) The air-fuel ratio is the ratio of the weight of air to the weight of gasoline. For example, a ratio of 14 to 1 means that 14 pounds of air are required for every pound of gasoline.

9. (A) A venturi is a constriction of a flow path. A reduction of pressure occurs in the venturi throat as the air moves through, the amount of reduction in a venturi of a given size depending upon the speed of the movement. Illustration (A) most nearly indicates the condition that would occur in a venturi in the throat of a

carburetor. The reduction of pressure in the venturi is used to cause the fuel to flow from the bowl into the air stream.

10. (D) An enriched mixture is required for a wide-open, full-power operation. The main jet in the carburetor provides for economical operation through the normal range of operations, but is not designed to provide the extra fuel required at full power. This extra fuel is supplied by the power enrichment system.

11. (A) The rapidity with which an upshift can be made in any unsynchronized transmission is limited by the time it takes for the freespinning clutch disc and countershaft assemblies to slow down to the more slowly rotating speed of the transmission mainshaft. Clashless sliding gear engagement is possible only at the moment these parts synchronize in speed. A clutch brake can be used to reduce this slow-down time and hold the apparatus momentum loss to a minimum.

12. (B) The primary advantage of disc brakes is the reduced brake fading and, consequently, a shorter stopping distance. Disc brakes are used only on the front wheels of some vehicles, while others have them on all wheels.

13. (A) Several protective devices are employed to regulate the voltage and current output of the generator (alternator) and to maintain a fully charged storage battery; the choice depends on the kind of electrical system. The most representative of these devices is the "three-unit" regulator. This regulator contains a cut-off relay, a current regulator, and a voltage regulator.

14. (C) The length of an aerial ladder is measured by a plumb line which extends from the top rung of the ladder to the ground, when the ladder is fully extended at its maximum elevation. The maximum elevation with most ladders is at an angle of approximately 85° from the ground.

15. (C) The tractor-trailer combination is more maneuverable than the other two. It can be taken into spots that are inaccessible to either the rear- or mid-ship mounted apparatus. Because it can be jackknifed, the tractor-trailer combination can also be positioned in a manner that provides more stability when the aerial ladder must be operated in other than an in-line position.

16. (C) It is essential that aerial ladders be as light as possible, yet be constructed in such a manner and of such material that they have sufficient strength when operated in either a supported or an unsupported position. The construction best suited for these operations is referred to as truss construction. This method of construction permits tension and compression stresses to be distributed for maximum strength and provides almost equal strength in either a tension or compression situation.

17. (A) Levers are usually self-centering, returning to neutral in the event the operator's hand slips off the control.

18. (C) The most stress is placed on the apparatus when the ladder is

fully extended in a cantilever position horizontal to the plane of the earth. Various stresses take place between this maximum stress position and the fully elevated position, and will shift as the ladder is rotated.

19. (A) Limitations on the use of aerial ladders depend on the angle of extension and whether the ladder is operated in a supported or unsupported position. A device provided on most aerial ladders with a length of 75 feet or more informs the apparatus operator of operating limitations. It is referred to as an inclinometer. The most common kind of inclinometer is mounted on the aerial ladder and is free to swing with the elevation of the ladder. The inclinometer indicates the maximum safe extention of the ladder at various angles when operated supported, unsupported, or as a water tower.

20. (C) Jackknifing denotes the turn of the tractor out of line with the trailer in order to provide better apparatus stability. The ideal jackknife is an outside jack which provides a 60° angle between the tractor and the trailer. In an outside jackknife, the ladder is raised away from the complement of the jackknife angle; for an inside jack, the ladder is raised toward the complement of the jackknife angle.

21. (C) The longest elevated platform in use is a 150-foot Firebird. This apparatus is manufactured by Calavar and is of the combination telescopic-articulated type.

22. (C) In addition to being identified by boom arrangement, elevated platforms are designated by effective working height. The effective working height is measured by a plumb line from the floor of the platform to the ground, when the platform is raised to its maximum working height. Elevated platforms are available in sizes ranging from 55 to 150 feet. The longer platforms are of the combination telescopic-articulated type.

23. (D) Horizontal reach is just as important, if not more important, than the working height. The horizontal reach of a 85-foot aerial ladder is approximately 80 feet while the reach of an 85-foot articulated elevated platform is about 44 feet.

24. (C) Platform-carrying capacity varies between 700 and 1000 pounds, depending on the manufacturer and the rated working height. This rating prevails at both the maximum height and the maximum reach; however, the rating decreases on some types when heavy streams are placed in operation.

25. (B) A deadman control is a device to which pressure must be applied before another system can be placed in operation. A deadman control is provided on elevated platforms for the operation of the boom system. The booms cannot be activated until pressure is applied to the deadman control, and the system will be deactivated if the pressure is released. This means that positive pressure must be maintained on the deadman control at all times to operate the boom system. The objective of this control is to stop operations in the event of the collapse of the operator.

Water Supply

Answers to questions in this section are based upon information contained in the text <u>Fire Department Hydraulics</u> (Eugene F. Mahoney, Allyn and Bacon, Inc., Boston, 1980). This information has been used with the expressed permission of Allyn and Bacon, Inc.

1. When considering water supply systems for a municipality, the Insurance Services Office Grading Schedule

   (A) gives more credit for gravity systems.
   (B) gives more credit for pumping systems.
   (C) makes no distinction between a gravity system and a
        pumping system.
   (D) Does not evaluate pumping systems.

2. The average daily consumption of water in cities in the United States is most nearly

   (A) 110 gallons per capita.
   (B) 150 gallons per capita.
   (C) 200 gallons per capita.
   (D) 250 gallons per capita.

3. The maximum daily consumption is the maximum amount of water used in a city during any twenty-four-hour period in

   (A) a year.
   (B) two years.
   (C) three years.
   (D) five years.

4. Water systems should be designed to supply the required fire flow to every section of the city while the domestic use is at the

   (A) average daily consumption rate.
   (B) maximum daily consumption rate.
   (C) peak hourly consumption rate.
   (D) peak three-hour consumption rate.

5. The minimum required fire flow for any area within a city for a single fire is

   (A) 250 gpm.
   (B) 500 gpm.
   (C) 750 gpm.
   (D) 1000 gpm.

6. A water supply system is considered adequate when it can deliver the required fire flow for the

   (A) required duration.
   (B) required duration while the domestic consumption is
        at the average daily rate.
   (C) required duration while the domestic consumption is
        at the maximum daily rate.

(D)   required duration while the domestic consumption is
at the peak hourly rate.

7.   A water system is considered to be reliable when it can supply the
required fire flow for the number of required hours, with domestic
daily consumption at the

(A)   average daily rate under certain emergency or unusual
conditions.
(B)   maximum daily rate at all times.
(C)   maximum daily rate under certain emergency or unusual
conditions.
(D)   peak hour rate under certain emergency or unusual
conditions.

8.   The minimum acceptable pressure on a water supply system is

(A)   10 psi.
(B)   16 psi.
(C)   20 psi.
(D)   25 psi.

9.   The normal range of pressures in the water supply system of most
cities is

(A)   35 to 45 psi.
(B)   45 to 55 psi.
(C)   55 to 65 psi.
(D)   65 to 75 psi.

10.   Primary feeders in water supply systems should be

(A)   at least 12 inches in size.
(B)   at least 18 inches in size.
(C)   at least 24 inches in size.
(D)   of sufficient size to deliver the required fire flow
to all built-up areas of a community, with the domes-
tic consumption at the maximum daily rate.

11.   It is recommended that secondary feeders be spaced not more than
how far apart?

(A)   1000 feet
(B)   1500 feet
(C)   2000 feet
(D)   3000 feet

12.   What is the minimum-size main recommended on all principal streets?

(A)   8-inch
(B)   10-inch
(C)   12-inch
(D)   14-inch

13. What is the minimum-size main recommended for residential areas?

    (A) 4-inch
    (B) 6-inch
    (C) 8-inch
    (D) 10-inch

14. Sufficient control valves should be installed in water systems so that a break in a pipe in the commercial area will not necessitate shutting down a pipe more than what length?

    (A) 500 feet
    (B) 800 feet
    (C) 1,000 feet
    (D) 1,200 feet

15. A standard fire hydrant is hydrostatically benched tested at

    (A) 150 psi.
    (B) 200 psi.
    (C) 300 psi.
    (D) 500 psi.

16. A standard fire hydrant should be capable of flowing a minimum of

    (A) 500 gpm.
    (B) 600 gpm.
    (C) 750 gpm.
    (D) 1000 gpm.

17. The California hydrant is a

    (A) dry barrel hydrant.
    (B) wet barrel hydrant.
    (C) flush type hydrant.
    (D) high-pressure hydrant.

18. The most commonly used type of hydrant is the

    (A) dry barrel hydrant.
    (B) wet barrel hydrant.
    (C) flush type hydrant.
    (D) high-pressure type hydrant.

19. Flush type hydrants are

    (A) dry barrel hydrants.
    (B) wet barrel hydrants.
    (C) either wet barrel or dry barrel hydrants.
    (D) not really hydrants.

20. Dry hydrants will most likely be found

    (A) in high-value districts.
    (B) in residential areas.
    (C) on private property.

(D)   in rural areas.

21.   The NFPA recommends that the bonnet and nozzle caps of hydrants able to provide flows of 500 to 1000 gpm be painted what color?

(A)   yellow
(B)   green
(C)   orange
(D)   red

22.   It is desirable that hydrants be not more than how far apart in any area of a city?

(A)   300 feet
(B)   500 feet
(C)   800 feet
(D)   1,000 feet

23.   From a fire protection standpoint, the ideal time for testing a water system is

(A)   between 8:00 A.M. and 4:00 P.M.
(B)   between 4:00 P.M. and 12:00 midnight.
(C)   between 12:00 midnight and 8:00 A.M.
(D)   when the domestic demand on the system is the highest.

24.   When the quantity of flow from a water system exceeds 1,000 gpm, the system should be tested for an accuracy with an error of not more than

(A)   10 gpm.
(B)   25 gpm.
(C)   50 gpm.
(D)   100 gpm.

25.   When testing a water system, a pitot blade is used to

(A)   determine flow pressure.
(B)   flow directly.
(C)   record pressure drops.
(D)   calculate static pressure.

Answers

1. (C) Although a gravity system is more reliable than a pumping system, the Insurance Services Office Grading Schedule makes no distinction between the two, as long as the pumping system is well-designed and equipped with adequate safeguards.

2. (B) The average daily consumption of water for a city is determined by dividing the total amount of water used in a one-year period by the number of days in the year. The average daily consumption of cities in the United States is approximately 140 to 150 gallons per capita.

3. (C) The maximum daily consumption is the maximum amount of water used in a city during any twenty-four-hour period in a three-year period. A day is not considered in determining the maximum daily consumption if an excessive amount of water is used in an unusual situation, such as refilling a reservoir after cleaning. The maximum daily consumption is normally about 150% of the average daily consumption.

4. (B) Water systems should be designed to supply the required fire flow to every section of the city while the domestic use is at the maximum daily consumption rate. The required fire flow is the amount of water needed for firefighting in order to confine a major fire to the buildings within a block or other group complex.

5. (B) The minimum required fire flow for any area within a city for a single fire is 500 gpm while the maximum is 12,000 gpm. The 500-gpm minimum is basically for noncongested areas of small dwellings.

6. (C) A water supply system is considered adequate when it can deliver the required fire flow for the required duration of hours while domestic consumption is at the maximum daily rate. The required duration of hours varies with the required fire flow.

7. (C) A water system is considered reliable when it can supply the required fire flow for the number of required hours, with the domestic daily consumption at the maximum daily rate under certain emergency or unusual conditions. By necessity, there must be a certain amount of duplication in a system for it to be considered reliable.

8. (A) If water mains are large and hydrants are well distributed and of the proper type and size, it is possible to take a minimum pressure as low as 10 psi.

9. (C) Although the minimum acceptable pressure can be as low as 10 psi, the normal pressure range in most cities is 65 psi to 75 psi. Pressures in this range are adequate for ordinary consumption in most buildings up to ten stories and are also sufficient to provide for sprinkler systems in buildings up to five stories.

10. (D) Primary feeders are large pipes that are used for moving water from the source of supply or storage area to the secondary feeders. The size of the primary feeders may vary from 60 inches in larger cities to 12 inches in smaller ones. Regardless of the size,

primary feeders should have a capacity sufficient to deliver the required fire flows to all built-up areas of a community, with domestic consumption at the maximum daily consumption rate.

11. (D) Secondary feeders are smaller than primary feeders, but larger than the distribution mains used in the grid system. Secondary feeders tie the grid system to the primary feeders to aid in concentrating the required fire flow at any point within the grid network. It is recommended that secondary feeders be spaced not more than 3,000 feet apart.

12. (C) Twelve-inch mains are recommended on all principal streets, with 8-inch mains cross-connected every 600 feet in business districts.

13. (B) The minimum-size mains recommended for residential areas is 6 inches, with these mains cross-connected at intervals of not more than 600 feet.

14. (A) It is recommended that sufficient valves be installed in water supply systems so that a break or failure of the system will not require shutting down a length of pipe more than 500 feet in commercial areas, or a quarter-mile for arterial mains.

15. (C) Standards for hydrants are prepared by the American Water Works Association. A standard hydrant is designed for a working pressure of 150 psi and is hydrostatically benched tested at 300 psi.

16. (B) A standard hydrant should be capable of flowing a minimum of 600 gpm with a pressure loss of not more than five pounds between the street main and the outlet.

17. (B) The wet barrel hydrant (sometimes referred to as the California type) is filled with water at all times, making water available at the hydrant outlet as soon as the hydrant valve is opened.

18. (A) The dry barrel hydrant is currently the most commonly used. On this hydrant, the valve controlling the water is located below the frost line; thus, the barrel of the hydrant is dry except when in use. At the base of the hydrant there is a drain valve, which permits the hydrant barrel to be drained after use. The drain valve is designed so that it closes as the main valve is opened, and opens as the main valve is closed.

19. (C) Flush type hydrants are those in which the outlets and the control valves are located below ground level. A metal plate is generally used to cover the cast-iron box or manhole in which the hydrant is located. The words "fire hydrant" might be found on the metal cover. Flush type hydrants may be either wet barrel or dry barrel, depending upon where they are installed.

20. (D) Dry hydrants are not connected to a positive pressure source of water. In essence, they are permanently installed hard-suction lines. A pipe extends from the hydrant to the body of water. A strainer is attached to the end of the piping. A pumper connects to the hydrant and proceeds to draft just as if a hard-suction line

had been lowered into the water. Such hydrants are found in rural areas or near piers and are so located that water may be drawn from pools, lakes, rivers, and such.

21. (C) The NFPA recommendations are that the barrel of a hydrant be painted chrome yellow, and that the bonnet and nozzle caps be painted as follows:

    For flows of 1,000 gpm or more   green
    For flows of 500 to 1,000 gpm   orange
    For flows of less than 500 gpm   red

22. (C) As a general rule, hydrants should be placed at each street intersection. When blocks are extremely long, there should be additional hydrants in the center of the block. It is desirable that hydrants be not more than 800 feet apart in any area of a city, and not more than 500 feet apart in built-up areas.

23. (D) From a fire protection standpoint, the ideal time for testing a water system is when the domestic demand on the system is the highest. Usually, this period is between 9:00 A.M. and 5:00 P.M., Monday through Friday, although it can vary.

24. (D) A high degree of accuracy is not required when testing the water system. The objective of the tests is to determine the number of fire streams obtainable from the system. A standard 2½-inch, hand-held line discharges 250 gpm. Fire flow tests need only be calculated to the nearest 50 gpm for total quantities of less than 1,000 gpm, and to the nearest 100 gpm for quantities in excess of 1,000 gpm.

25. (A) Pitot blades are used to determine the flowing pressure from a hydrant or nozzle stream. It is important that they be inserted in the flowing stream without causing excessive wobble or spray. Readings should be as accurate as possible. Usually, either a pressure-gage snubber or an air chamber is attached to the blade to remove any gage needle fluctuations. The pressure snubber is the slower of the two but gives better results.

V

FIRE PREVENTION

Practices and Principles

1. Regardless of how a law may look on the books, for 90% of the people that a fire inspector advises, the law is

   (A) irrelevant.
   (B) what the inspector says it is.
   (C) an arbitrary imposition on their liberty.
   (D) only for criminals and mental defectives.

2. Floor or wall openings sometimes prevent the banking up of heated air. This condition, with respect to sprinklers, is considered

   (A) advantageous.
   (B) unimportant.
   (C) detrimental.
   (D) good ventilation.

3. Of the following, which has been the most frequent factor contributing to conflagrations in the United States in the last twenty-five years?

   (A) high winds
   (B) lack of exposure protection
   (C) wood-shingle roofs
   (D) inadequate water distribution system

4. Which of the following is not considered one of the essential three "E"s of a fire prevention program?

   (A) engineering
   (B) enlistment
   (C) education
   (D) enforcement

5. For the most effective results in conducting a Fire Prevention Week campaign, it would be desirable to emphasize fire prevention

   (A) in its broader community aspects.
   (B) as a means of lowering insurance rates.
   (C) as it applies to the individuals' own homes.

(D) as a means of lowering the operating costs of the fire department.

6. To best analyze the fire prevention and protection problems in a certain section of the city, the most basic thing that it is necessary to know is

(A) the number of fire companies in the area.
(B) the structural and occupancy data of the area.
(C) the number of people living in the area.
(D) the available water supply for the area.

7. The primary objective of a fire prevention inspection should be

(A) the elimination of fire hazards.
(B) public relations.
(C) pre-fire planning.
(D) to gather information.

8. In an inspection of a medical laboratory, the fire officer found that ethyl ether was stored in a modern domestic household refrigerator to reduce possible vaporization. Of the following, which is the most likely reason for considering this method of storage inadvisable?

(A) Desired temperature ranges for such storage are not normally available in domestic refrigerators.
(B) Released gases or vapors may escape without detection in such storage spaces.
(C) Explosion vents are usually lacking in the refrigerated storage space.
(D) Exposed electrical parts of the refrigerator are usually located in the storage space.

9. "It is good practice to so install heating devices that under conditions of maximum heat (long continued exposure) they will not cause the temperature of exposed woodwork to exceed 160°F." This practice is

(A) correct because of the possibility that wood and other combustible materials, after long continued exposure to relatively moderate heat, may ignite at temperatures far below their usual ignition temperature.
(B) not correct because no wood in ordinary use will ignite at a temperature of less than 400°F and consequently the requirement is needlessly severe.
(C) correct because oxidation proceeds much more rapidly at higher temperatures.
(D) not correct because oxidation proceeds much more slowly at higher temperatures.

10. The difference between flammable liquids and combustible liquids is expressed in terms of

(A)   flash point only.
(B)   flash point and vapor pressure.
(C)   flash point and boiling temperature.
(D)   flash point, vapor pressure, and boiling temperature.

11.   Fire prevention laws can most effectively be enforced by means of

(A)   prompt issuance of citations.
(B)   stiff penalties for noncompliance.
(C)   police personnel instead of fire personnel.
(D)   public education.

12.   The real problem in fire prevention and protection in the presence of radioactive isotopes is that the radioactivity of an element or compound

(A)   is not diminished by changes in the compound caused by fire or explosion.
(B)   increases directly with the intensity of fires occuring in the establishments using them.
(C)   causes a compound to vaporize, melt, or oxidize more readily.
(D)   is reduced significantly if control by means of suitable sprays or extinguishers is immediately applied.

13.   From the standpoint of fire prevention, air-conditioning and air-blower systems are of concern mainly because they

(A)   provide a means for the spread of fire through the building served.
(B)   severely limit adequate ventilation in case of fire.
(C)   intensify fires from other sources by providing abnormally large amounts of air.
(D)   characteristically accumulate hazardous quantities of dust and lint which are subject to spontaneous ignition.

14.   In fire prevention planning, a decision tree is concerned with

(A)   the control of explosives.
(B)   the control of hazardous chemicals.
(C)   building fire safety.
(D)   radioactive materials.

15.   The chief objective of pre-fire planning inspections performed by firefighters in the fire department is to

(A)   discover and remedy petty violations of fire prevention ordinances and bad prevention practices.
(B)   keep the public informed and aware of fire hazards.
(C)   maintain a direct relationship between the firefighters and the public and to keep the public aware that the fire department is on its toes.
(D)   make possible more efficient and intelligent firefighting.

16. The fire load computation of a building indicates, for the most part, the

    (A) risk of a fire breaking out.
    (B) rate at which a fire is likely to grow.
    (C) combustibility of the various parts of the building rather than its contents.
    (D) maximum fire stress to which the building might be subjected.

17. What, most nearly, is the minimum width of exit usually required for a single file of persons?

    (A) 15 to 17 inches
    (B) 18 to 20 inches
    (C) 21 to 23 inches
    (D) 24 to 27 inches

18. According to NFPA, a fire death is a fire casualty which is fatal within

    (A) twenty-four hours of the fire.
    (B) one week of the fire.
    (C) one month of the fire.
    (D) one year of the fire.

19. The underlying reason for routine periodic and frequent fire prevention inspection is

    (A) that occupants, hazards, and code compliances may vary considerably in a given building over a short period.
    (B) the need for favorable public opinion.
    (C) that a large city usually has many new buildings being constructed.
    (D) that most individuals continually and consciously try to evade the fire regulations.

20. What public attitude toward fire prevention is most difficult to overcome?

    (A) maliciousness
    (B) indifference
    (C) laziness
    (D) obstinacy

21. The most prolific structural cause of fire spread in non-fireproof hotels, other than of unprotected vertical openings, is

    (A) inadequate auxiliary appliances and alarm systems.
    (B) unprotected horizontal openings.
    (C) poorly protected storage of paints, oils, wastes, and volatile mixtures.
    (D) inadequately protected linen and furniture storage rooms.

22. The fire in the United States that caused the greatest loss of life occurred

    (A) aboard a ship
    (B) in a theater.
    (C) in a school.
    (D) in an industrial plant.

23. The simplest and most feasible method of avoiding the overheating of woodwork near any high temperature heating appliance is by

    (A) filling the intervening space with insulating material.
    (B) covering the woodwork with sheet metal.
    (C) providing an air space between the woodwork and the appliance.
    (D) covering the woodwork with asbestos sheets.

24. In a large city, the number of fire prevention inspections in each of the various sections of the city should usually be proportional to the

    (A) number of inhabitants of the section.
    (B) size of the section.
    (C) fire hazards in the section.
    (D) assessed value of property in the section.

25. The combined use of inspections, period reports of activities, follow-up procedures, special reports from subordinates, and a rating system constitute a system of

    (A) coordination.
    (B) command.
    (C) control.
    (D) representation.

26. A fire inspector should know that an inert gas is one that

    (A) will not support combustion.
    (B) will burn.
    (C) will explode on contact with certain chemicals.
    (D) has been converted to a liquid.

27. The materials of which a building is constructed and the opportunity for the spread of fire are important, but the biggest single hazard is usually that of

    (A) occupancy.
    (B) location.
    (C) fire protective measures.
    (D) construction.

28. The first objective of all fire prevention is

    (A) safeguarding life against fire.
    (B) reducing insurance rates.
    (C) preventing property damage.

(D)   confining fire to a limited area.

29.   Upon arrival at a commercial building for the purpose of making a
fire prevention inspection, it is best that the inspector

(A)   immediately contact the manager.
(B)   take a good look outside the building before going
        inside.
(C)   take a quick walk around the inside of the building
        before contacting the manager.
(D)   talk to several employees to find out if any fire
        regulations have been violated.

30.   "More ink has been spilled on the item of smoking as a cause of
fire than on any other, but the total result has been negligible."
Which of the following best accounts for this situation?

(A)   Smoking is generally an automatic act performed un-
        thinkingly.
(B)   Truly effective facilities for elimination of smoking
        hazards are exceedingly cumbersome or expensive.
(C)   Most people regard smoking as a personal prerogative
        and resent control measures.
(D)   Smoking is practiced by many individuals with defec-
        tive intelligence and social attitudes.

Answers

1. (B) Most people are not familiar with fire prevention laws, fire codes, or building codes. However, they seem to place a tremendous amount of faith in the fire inspector. People will accept what the inspector says as being the law. Fire inspectors must be extremely careful in dealing with the public so that they do not betray this trust. They must be honest with people in making recommendations for correction of hazards or for the purpose of enhancing fire safety.

2. (C) In order for a sprinkler head to extinguish or control a fire, it is necessary that heat from the fire reach the head directly above the fire location. Any condition that restricts the build-up of heat over the fire is detrimental to the successful operation of the sprinkler system. Floor or wall openings sometimes contribute to this detrimental effect by creating drafts that carry the heat from the fire away from the fire location.

3. (A) For many years wood-shingle roofs were the primary contributor to conflagrations; however, with more and more jurisdictions enacting laws outlawing these roofs, it has lost its place as the number one contributor. The number one contributor during the past twenty-five years has been high winds, or those with a velocity in excess of 30 mph. The number two contributor has been inadequate water distribution systems.

4. (B) The various methods used by a fire department in a fire prevention program can be summed up by engineering, education, and enforcement. Engineering involves the methods used for building fire protection and safety into a structure. Education is concerned with instructing the public in the hazards and dangers of fire, and enforcement is the legal means used to insure that the public complies with the fire laws.

5. (C) Unjustified as it may be, people are usually more concerned with their own personal safety than they are with that of their neighbors. A fire prevention campaign is designed to sell fire prevention, and it is much easier to sell something to a person who can see how it directly relates to him or her. Consequently, it is desirable to emphasize fire prevention as it applies to an individual's own home or place of business.

6. (B) To analyze the fire protection problem of a portion of the city, it is necessary to know the structural and occupancy data for the area, or in other words, "what there is to burn." To carry this a step farther, to analyze the fire protection problem of a particular building, it is necessary to know the structural and occupancy data for the building.

7. (A) After considering life safety, the primary objective of a fire prevention inspection should be the discovery and elimination of fire hazards. In most cases, the elimination of fire hazards is based upon the enforcement of the fire code; however, every effort should be made to sell the importance of fire prevention to the occupant.

8. (D) Ethers are highly flammable and have been known to ignite from an energy source such as a hot plate. It is extremely important to keep ethers isolated from all sources of ignition. The exposed electrical parts in refrigerator storage space could serve as an ignition source in the event of escape of the ether gases.

9. (A) Various woods have different ignition temperatures; however, the commonly reported ignition temperature for wood is in the order of 392°F. Wood, when subjected to constant temperature sources over a long period of time, will undergo chemical changes with a resultant change in the ignition temperature. It has been recommended that the highest temperature to which wood should be continuously subjected without risk of ignition is 212°F. Holding the maximum temperature to 160°F maintains safety.

10. (B) A flammable liquid is any liquid having a flash point below 100°F and having a vapor pressure not exceeding 40 psi at 100°F. A combustible liquid is a liquid having a flash point above 100°F.

11. (D) Most fire prevention laws have been written to protect the public. In general, people prefer to live and work in a safe environment. If people understand the purpose of laws in regard to their own and others' safety, it is not normally too difficult to achieve compliance. Public education is one of the primary tools (one of the three "E"s) that should be used by fire personnel to secure fire safe conditions.

12. (A) The fire and explosive hazard of a material is the same regardless of whether or not the material is radioactive. The problem involved with handling emergencies of radioactive material is that the hazard of the radiation is not eliminated as the material burns. Radiation during the firefighting phase continues and will cause damage to living tissue if firefighters are not adequately protected. Efforts must be made to ensure that any contained radioactive material is not released during firefighting operations, if possible.

13. (A) Although air-conditioning and air-blower systems contribute to the daily comfort of individuals, they do increase the problem to firefighters. From a fire prevention standpoint, these systems must be evaluated in relation to how they will contribute to the spread of fire. From a firefighting standpoint, firefighters must be aware of these systems both in terms of fire spread and how they can be used in fire control.

14. (C) The decision tree is a NFPA concept concerned with the elements that must be considered when evaluating building fire safety and the interrelationship of these elements. Successful use of the tree depends upon how completely each level of elements is satisfied.

15. (D) The objective of a pre-fire planning inspection is to gather as much information as possible to help fight the fire before it occurs. During the inspection, firefighters should note the methods of entry into the building, how many people are inside, the best method of evacuation, the location of hazards, the location

of utility shutoffs, ventilation problems, the amount of hose required to reach various sections of the building, and many other facts. Gathering this information will assist in more intelligently fighting the fire when it occurs.

16. (D) Fire load is a term used to denote the amount of combustible material expected in a given fire area. The amount of heat that will be given off during a fire can be calculated based upon the calorific value of the materials involved. While the severity of a fire will be determined by the type of material burning and the rate of burning, the overall fire stress that can be expected is based upon the fire load, or what there is to burn.

17. (C) The NFPA Life Safety Code establishes 22 inches as the recommended exit width. This recommendation has been adopted by many communities for use as a basis for exit requirements in the establishment of occupancy loads.

18. (D) A fire death is a fire casualty that is fatal or becomes fatal within one year of the fire. A fire casualty is a person receiving an injury or death resulting from a fire.

19. (A) There is an old saying that there is nothing constant but change. This is particularly true for fire prevention. An inspector can inspect a building one day and give the occupant a clean bill of health, yet find conditions in the same building the next day to be extremely hazardous. The rapid change of conditions particularly applies to housekeeping activities but is also important in regard to storage.

20. (B) People in the United States strongly believe that it can happen to their neighbors, but it can't happen to them. This indifference to the hazard of fire is one of the primary reasons for our tremendous fire loss. It is not unusual for an inspector trying to sell fire prevention to get an answer such as, "I've been in business here for twenty years, and I've never had a fire yet."

21. (B) There are two primary structural defects contributing to fire spread in non-fireproof hotels. One is unprotected vertical openings, and the other is unprotected horizontal openings. Although unprotected vertical openings is the biggest contributor to large fires in older hotels, unprotected horizontal openings certainly play a major role.

22. (A) The largest loss of life to fire in the United States occurred aboard the excursion steamer General Slocum in 1904. The loss of life was 1,030 in this fire. The fire causing the second largest loss of life occurred in the Iroquois Theater in Chicago in 1903.

23. (C) Providing adequate space between a heating unit and nearby woodwork is the simplest and most feasible method to avoid overheating. It must be remembered that the amount of heat reaching an exposed object is inversely related to the square of the distance. An object four feet from a heat source will receive only one-fourth the heat of an object two feet from the source.

24. (C) How often inspections are made should be decided according to the fire hazards. If possible, some occupancies should be inspected on a daily basis in order to maintain the premises in a safe condition. Of course, in general, this is impractical and would not be accepted by the public. However, target hazards should be inspected at least four times a year, and more often as the situation demands.

25. (C) Control, as a management term, refers to the establishment of methods to insure that those things that should be done are done. Some control methods used in the management of fire prevention activities are inspections, reports, follow-up procedures, and rating systems.

26. (A) An inert gas is one that will not support combustion. Some of the common inert gases are nitrogen, carbon dioxide, helium, and sulfur dioxide.

27. (A) The occupancy of a structure is usually considered the single factor contributing most to the fire hazard. Using an extreme example, it is easy to see that the hazard in a fireworks manufacturing plant is much greater than it is in a school. It is also easy to see why a woodworking shop would normally be more hazardous than a motel.

28. (A) The first objective of fire protection is the safeguarding of life against fire. This is true of both fire prevention and firefighting. The first thought of a fire inspector should be to make the building safe for the occupants, and the first thought of a firefighter should be the rescue of those who might be injured or killed as a result of the fire.

29. (B) When arriving at the building, it is important that the inspector take a good look outside the building before going inside. This will give the inspector an opportunity to observe exposure hazards, fire escapes, fire department inlets, the height of the building, and other factors that will assist in the actual inspection of the inside of the building.

30. (A) Fires caused by smoking are generally due to carelessness. Discarded cigarettes continue to burn at a temperature sufficiently high to ignite combustibles. Smoking is such an automatic habit that many people discard cigarettes without realizing what they have done.

Fire Chemistry and Behavior

1. When firefighters direct water on an exposure, they are doing so to keep the exposure below its

    (A) ignition temperature.
    (B) fire point.
    (C) explosive range.
    (D) flash point.

2. Suppose a fire is burning in a two-story residence. Building A is 20 feet away from the fire, while Building B is 80 feet distant from the fire. The amount of radiated heat received by Building A as compared with Building B may be expected to be most nearly

    (A) four times as great.
    (B) six times as great.
    (C) twelve times as great.
    (D) sixteen times as great.

3. Volatile means, most nearly,

    (A) readily flammable.
    (B) capable of being easily evaporated.
    (C) highly explosive.
    (D) easily detectable by the sense of smell.

4. Fire can be extinguished only by

    (A) removing the fuel and excluding oxygen.
    (B) excluding oxygen and reducing the temperature.
    (C) removing the fuel and reducing the temperature.
    (D) elimination of any one of the factors mentioned.

5. Prevention of the generation of static electricity is considered to be

    (A) always possible.
    (B) usually possible.
    (C) rarely possible.
    (D) impossible.

6. The explosive limits of a flammable liquid are 2.5 to 8.3. The concentration of vapors in air of this liquid is 2.1%. It can best be said that the vapors

    (A) are likely to burn if exposed to a flame.
    (B) are too lean to burn.
    (C) are too rich to burn.
    (D) will ignite only if exposed to a heat source above the ignition temperature of the liquid.

7. The ratio of the density of a material to the density of some standard substance is the

    (A) diffusion rate.

(B)   specific gravity.
(C)   miscibility.
(D)   explosive range.

8.   Which of the following statements is a description of the transmission of heat by radiation?

(A)   Heat is traveling through space evenly in all directions.
(B)   Heat is traveling through the motion of heated matter.
(C)   Heat is traveling through matter without visible movement of matter.
(D)   Heat is traveling by heated air, gases, or water.

9.   In relation to flammable liquids, the hazard of flammable gases is generally

(A)   opposite.
(B)   dissimilar.
(C)   similar.
(D)   identical.

10.   Reid is specifically identified with

(A)   combustibility.
(B)   ignition point.
(C)   flash point.
(D)   vapor pressure.

11.   It can best be said that a flammable liquid has

(A)   one flash point.
(B)   two flash points.
(C)   three flash points.
(D)   four flash points.

12.   The specific heat of a substance is best defined as the

(A)   rate at which the substance will absorb heat as compared to water.
(B)   number of BTUs required to raise the temperature of a pound of a substance one degree Fahrenheit.
(C)   number of BTUs of heat produced by the combustion of one pound of the substance.
(D)   heat-resistive quality of the substance as compared to Fuller's earth.

13.   Which of the following properties is most important in determining the degree to which vapor travel is a hazard of a flammable liquid?

(A)   flash point
(B)   latent heat
(C)   vapor pressure
(D)   vapor density

14. Explosions of vapor-air mixtures inside tanks containing flammable liquids are most frequent when the tank

    (A) is full.
    (B) has just been emptied.
    (C) is being filled.
    (D) is half full.

15. "There is little if any chemical reaction between most combustibles and the oxygen of the air at ordinary temperatures, but, when heated sufficiently, reaction begins and continues until the combustible reaches a temperature at which the reaction becomes self-sustaining." This temperature level is known as the

    (A) flash point.
    (B) critical temperature.
    (C) ignition point.
    (D) explosive point.

16. Heat of combustion most nearly refers to the

    (A) rate at which oxygen is consumed during the burning process.
    (B) amount of heat necessary to initiate combustion.
    (C) amount of heat released during the combustion process.
    (D) intensity with which a fire burns.

17. The amount of heat required to raise the temperature of one pound of water one degree Fahrenheit is defined as the

    (A) gram calorie.
    (B) specific heat of a substance.
    (C) latent heat unit.
    (D) British Thermal Unit.

18. Which of the following relationships between flash point and the storage temperature of flammable liquids is most likely to produce a vapor-air mixture above the liquid which is within flammable limits?

    (A) flash point substantially higher than storage temperature
    (B) flash point substantially lower than storage temperature
    (C) flash point approximately the same as storage temperature

19. A chemical compound which, when added to plain water in proper quantities, materially reduces the surface tension of plain water and increases its penetrating, spreading, and emulsifying ability, is

    (A) a drying agent.
    (B) a wetting agent.
    (C) a reducing agent.
    (D) an agitating agent.

20. The theory of the operation of a diesel engine is most closely associated with

    (A) frictional heat.
    (B) heat of solution.
    (C) heat of combustion.
    (D) heat of compression.

21. "The first five minutes are the most important." Which of the following scientific principles gives the most valid support to this firefighting axiom?

    (A) Energy can be neither created nor destroyed.
    (B) Mass and energy are different forms of the same thing.
    (C) For every action there is an equal and opposite reaction.
    (D) Heat accelerates chemical reaction.

22. The principal fire hazard from flammable liquids lies in the

    (A) low temperature at which they will ignite.
    (B) possibility of explosion while in the liquid state.
    (C) evaporation of the liquid when exposed to air or heat.
    (D) fact that the liquid itself will burn.

23. The occurrence of unusually high percentages of carbon monoxide at smoldering fires, such as furniture factories, would probably be primarily due to

    (A) unusually high temperatures.
    (B) the similarity of materials in most cases.
    (C) an oxygen deficiency.
    (D) unusually low temperatures.

24. The intensity of a fire depends upon the

    (A) temperature of the burning process.
    (B) material burning.
    (C) relative humidity of the air.
    (D) rate at which oxygen is supplied.

25. In order that ordinary combustible materials will no longer burn, it is necessary to reduce the oxygen in the air reaching the fire to what percentage?

    (A) 2%
    (B) 5%
    (C) 10%
    (D) 15%

26. The flash point of a liquid is which of the following?

    (A) the temperature at which a liquid boils
    (B) the temperature required to start actual combustion of the liquid

(C)   the temperature at which an ignitable vapor is
given off by the liquid

(D)   the temperature at which there is danger of a flash
from the liquid

27.   For each degree Fahrenheit increase in temperature, with constant
pressure, what, most nearly, is the increase in the volume of a
gas?

(A)   1/460
(B)   1/273
(C)   1/212
(D)   1/100

28.   Which of the following is not one of the four compounds found in
most ordinary combustible solids?

(A)   carbon
(B)   hydrogen
(C)   nitrogen
(D)   chlorine

29.   "The inherent characteristics of some materials are such that, under certain conditions of moisture and ventilation, an increase in
temperature without the material being exposed to external sources
of heat will create a hazard." Which of the following refers most
accurately to this phenomenon?

(A)   galvanic action
(B)   evaporation
(C)   dehydration
(D)   spontaneous heating

30.   "The amount of radiation reaching an exposed object varies directly as the square of the distance." This statement is accurate for

(A)   atomic radiation but not for heat radiation.
(B)   heat radiation but not for atomic radiation.
(C)   both atomic and heat radiation.
(D)   neither atomic nor heat radiation.

31.   The fire point is closest to the

(A)   ignition temperature.
(B)   flash point.
(C)   heat of combustion.
(D)   heat of compression.

32.   The phrase, "the ratio of the weight of any volume of a liquid to
the weight of an equal volume of water" most nearly means

(A)   molecular structure.
(B)   specific gravity.
(C)   weight.
(D)   vapor density.

33. A gallon of water, which will absorb 1,250 BTUs in the process of raising its temperature from 62°F to 212°F, will absorb, in addition, about how many times as many BTUs as it vaporizes and turns to steam?

    (A) the same
    (B) twice as much
    (C) three times as much
    (D) six times as much

34. Vapors are being given off by three containers sitting on a table. Vapor A has a vapor density of 0.8, vapor B a vapor density of 1.1, and vapor C a vapor density of 2.5. Which vapor would most likely be found at the floor level?

    (A) vapor A
    (B) vapor B
    (C) vapor C
    (D) none of them

35. With which of the following is boiling point most closely associated?

    (A) vapor pressure--atmospheric pressure
    (B) flash point--fire point
    (C) vapor pressure--flash point
    (D) atmospheric pressure--fire point

Answers

1. (A) Ignition temperature has been defined as the minimum temperature to which a substance in air must be heated in order to initiate or cause self-sustained combustion independent of the heating or heated element. Raising the temperature of a fuel above its ignition temperature when a sufficient amount of oxygen is available completes the fire triangle, resulting in the material breaking into flame. Playing water on an exposure is done for the purpose of keeping it from igniting.

2. (D) Radiant heat travels in waves at the speed of light. The amount of radiant heat reaching an exposed body varies inversely as the square of the distance separating the source from the body. As an example, in the problem, building B is four times farther from the heat source than building A. Therefore, the amount of heat reaching Building B as compared with that reaching building A would be 4^2 = 16 or one-sixteenth of that reaching building A.

3. (B) Webster defines volatile as "readily vaporizable at a relatively low temperature." Consequently, a volatile liquid is one that can be vaporized at a relatively low temperature.

4. (D) This question is based upon the theory of the fire triangle. According to this theory, fire can be extinguished by removing the heat, fuel, or oxygen. Consequently, answers (A), (B), and (C) would result in the extinguishment of the fire. It should be remembered, however, that fire can also be extinguished by breaking the chain reaction. The chain reaction added to the fire triangle forms the fire tetrahedron.

5. (D) The generation of static electricity cannot be prevented; however, static electricity can be controlled to prevent a hazardous situation from developing. One means is to bleed off the charge before sparking potentials are attained. Other means used to reduce the hazard are humification, bonding, and grounding.

6. (B) Vapors from a flammable liquid will burn if the concentration of air is within their explosive limits. Concentrations below the lower explosive limit are said to be too lean to burn, while those above the upper explosive limit are said to be too rich to burn. In this question, a 2.1 concentration is below the lower explosive limit of 2.5; therefore, the vapors are too lean to burn.

7. (B) Webster defines specific gravity as, "the ratio of the density of a substance to the density of another substance (as pure water or hydrogen) taken as a standard when both densities are obtained by weighing in air." Water is generally used as a standard; therefore, specific gravity is the weight of a volume of material when compared with an equal volume of water.

8. (A) Radiated heat travels in waves in all directions from a burning body. The heat travels at the speed of light. In contact with a material, it is either absorbed, reflected, or transmitted through the body.

9.   (C)   It should be remembered that it is not the flammable liquid that is the hazard but the vapors given off by the flammable liquid. The hazards of these vapors are similar to those of flammable gases.

10.   (D)   The Reid method determines the vapor pressures of petroleum liquids.   It is conducted in accordance with the ASTM Standard D-323 and gives the psi at 100 F.

11.   (B)   Contrary to most thinking, each flammable liquid has two flash points, the lower flash point and the upper flash point.   The lower flash point is given in most charts of hazardous materials.   It is the temperature at which sufficient vapors are given off to form a flammable mixture near the surface of the liquid.   This occurs at the lower explosive limit.   If an adequate source of ignition is available at this time, a flash will occur across the surface of the liquid and then go out.   The upper flash point is the temperature at which the concentration of vapors reaches the point where flame will no longer propagate.   This is known as the upper flash point and occurs at the upper explosive limit.   The upper and lower flash points change with a change in environmental pressure.

12.   (B)   Specific heat is the heat capacity of a substance.   It is defined as the number of BTUs required to raise the temperature of a pound of the substance one degree Fahrenheit.   One of the reasons that water is so effective in firefighting is that its specific heat is higher than other substances.

13.   (D)   Vapor density is an important factor when considering how far the vapors from a flammable liquid or LGP will travel to find a source of ignition.   As an example, butane has a vapor density of 2.0.   The vapors will collect at ground level and if winds or currents are not available to disperse them, the vapors will travel long distances.   This greatly increases the possibility of the vapors finding a source of ignition.

14.   (B)   In order to ignite a vapor-air mixture, the mixture must be present within its explosive range.   Most flammable liquids have flash points below the ambient temperature; consequently, sufficient vapors are being given off to enter the explosive range at ambient temperature.   Because of this, vapors within a container that is full or partially full are too rich to burn.   When the container has just been emptied, the vapor-air mixture within the tank is most likely within the explosive range and will ignite if a source of ignition is present.

15.   (C)   The reaction referred to in this question is spontaneous heating.   When spontaneous heating reaches the point where the reaction becomes self-sustaining, the material will break into flame.   This is known as the ignition point.   Incidentally, the critical temperature refers to the temperature above which a material can exist only in the gaseous state.

16.   (C)   Burning has been defined as rapid oxidation accompanied by heat and light.   Complete oxidation results in a substance being converted to carbon dioxide and water.   The amount of heat released during this oxidation process is referred to as the heat of

combustion. In plain language, it is the amount of heat given off when a substance burns.

17. (D) The British Thermal Unit (BTU) is a unit of measurement of heat. By definition, it is the amount of heat required to raise the temperature of one pound of water one degree Fahrenheit (measured at 60°F).

18. (C) Flash point is the temperature at which a flammable liquid will first give off sufficient vapors to enter the explosive range. If the flash point is substantially higher than the storage temperature, the vapors given off will be insufficient to burn. If the flash point is substantially lower than the storage temperature, sufficient vapors to burn will have been given off for a long time, and, consequently, the vapor-air mixture directly above the liquid will probably be too rich to burn. When the flash point and storage temperature are approximately the same, sufficient vapors to burn will just begin to be given off.

19. (B) Water has a relatively high surface tension. When fire burrows down in certain materials, such as a cotton mattress, the water will not penetrate. A wetting agent is an additive that reduces the surface tension of the water and allows it to penetrate for extinguishment. When these additives are used, the water is generally referred to as wet water.

20. (D) Heat is released when a gas is compressed. This is known as the heat of compression. The compression of air on the compression stroke of a diesel engine results in a release of heat. The amount of heat released elevates the temperature within the cylinder to a point above the ignition temperature of diesel fuel. At the proper time, an oil spray is injected into the cylinder. The heat within the cylinder is sufficient to cause immediate ignition of the oil.

21. (D) As a material burns, a certain amount of the heat produced by the burning feeds back into the material. This, in turn, accelerates the chemical reaction involved in the burning process, which increases the amount of heat given off. This feedback and increased chemical reaction is responsible for the rapid acceleration of the fire.

22. (C) One thought should be kept in mind. Flammable liquids do not burn. It is the vapors given off from flammable liquids that burn. Consequently, the danger of a flammable liquid is that sufficient vapors are given off to enter the explosive limits. One of the principal fire hazards of flammable liquids is that evaporation (vapors given off) of the liquid will occur when the liquid is exposed to heat or air.

23. (C) The product of complete combustion is CO_2 (carbon dioxide). The product of incomplete combustion is CO (carbon monoxide). Basically, when sufficient oxygen is present, each carbon atom from the burning material will combine with two oxygen atoms to form carbon dioxide. When there is an oxygen deficiency, each carbon atom will combine with only one oxygen atom. Therefore, the production of carbon monoxide by a burning fire is the result of an

oxygen deficiency.

24.   (D)   The intensity of a fire depends upon the rate at which oxygen is supplied. Normal burning, if there is such a thing, occurs when the oxygen content of the air is approximately 21%. As the oxygen content diminishes, the fire starts to die down and will eventually smolder and may even go out. Complete extinguishment will probably occur when the oxygen supply reaches 13%. On the other hand, if oxygen is supplied at a rate exceeding 21%, a definite increase in the intensity of the fire will occur. A good example is to apply pure oxygen to a burning cigarette. The entire cigarette will be consumed almost instantaneously.

25.   (D)   The 15% figure is an approximation. Some authorities say 16%; others say some form of combustion will take place in a concentration as low as 13%. It is important to recognize, however, that the human body and fire require about the same amount of oxygen to live. If there is insufficient oxygen for a fire to burn, there is insufficient oxygen for a person to live in the same atmosphere.

26.   (C)   The flash point of a liquid is the temperature at which sufficient vapors are given off to first enter the explosive range. If a constant spark is produced above a flammable liquid in an open container, when the flash point is reached, a flash will occur across the surface of the liquid and then go out. If heating of the liquid is continued, at a few degrees higher, a flash will occur across the surface of the liquid and continue to burn. This is known as the fire point.

27.   (A)   The answer to this question is based upon Charles's Law which states that the volume of a given mass of gas is directly proportional to the absolute temperature if the pressure is kept constant. Absolute zero temperature is $-459^{\circ}F$. It therefore follows that the volume of a gas will increase approximately 1/460 when the temperature changes from zero to one degree Fahrenheit if the pressure is kept constant. The absolute temperature when the reading is one degree Fahrenheit is 460 ($-459 + 1$).

28.   (D)   Although other compounds are present in ordinary combustible solids, most are composed of carbon, hydrogen, nitrogen, and oxygen. When a fire burns freely, the oxygen in the air combines with the carbon in the solid to form carbon dioxide, and with the hydrogen to form water vapor. Carbon dioxide is composed of two atoms of oxygen and one atom of carbon. When there is insufficient oxygen in the air, only one atom of oxygen combines with the carbon atom to form carbon monoxide, a deadly gas.

29.   (D)   Spontaneous heating, by definition, is the process of increasing the temperature of a material without taking heat from its surrounding. There are only a few basic causes of spontaneous heating; however, the conditions under which these few causes operate are many.

30.   (D)   The amount of radiation reaching an exposed object does not vary directly as the square of the distance but rather inversely as the square of the distance. Varying directly as the square of

the distance would mean that the farther from the source of the heat, the greater the radiation. The principle, however, of varying inversely as the square of the distance applies both to atomic and heat radiation.

31. (B) When an ignition source above the ignition temperature of a flammable liquid is present as the liquid is heated, a point will be reached in an open container where a flash occurs across the surface of the liquid and then goes out. This is known as the flash point. It occurs when the vapors given off by the liquid first enter the explosive range. If heating of the liquid continues, with the source of ignition remaining available, a point will be reached where a flash occurs across the surface of the liquid and continues to burn. This is the fire point. The fire point is a few degrees above the flash point.

32. (B) Specific gravity of a liquid is the ratio of the weight of a volume of the liquid to the weight of an equal volume of water. Water is therefore used as the standard and taken as 1.0. Those liquids having a specific gravity less than 1.0 will float on water (if not miscible). Water will float on liquids having a specific gravity greater than 1.0. With minor exceptions, flammable liquids have specific gravities less than 1.0.

33. (D) A BTU is the amount of heat required to raise one pound of water one degree Fahrenheit. To raise one pound of water from 62°F to 212°F would require 150 BTUs (212 - 62). As water weighs 8.33 pounds per gallon, it would take 1,245 BTUs (8.33 X 150) to raise the temperature of one gallon of water from 62°F to 212°F. However, to change one gallon of water from a liquid to steam requires approximately 8,082 BTUs (8.33 X 970.3). Consequently, it requires about six and one-half times as much heat to change water to steam as it does to raise its temperature from 62°F to 212°F (8082 divided by 1,245).

34. (C) Vapor density is the weight of a volume of a vapor (or gas) as compared with the weight of an equal volume of air (free from water vapor and carbon dioxide). Those vapors or gases having a vapor density of less than 1.0 are lighter than air while those having a vapor density of more than 1.0 are heavier than air. All other conditions being equal, the vapor or gas with the highest vapor density would most likely collect near the floor. Those with a vapor density of less than 1.0 would tend to move toward the ceiling while those with a vapor density near 1.0 would tend to concentrate near the level of disposal.

35. (A) As a liquid is heated, it will eventually reach a point where it can no longer be heated and still remain a liquid. The temperature at which this takes place is one where the vapor pressure and the atmospheric pressure are in equilibrium. This temperature is referred to as the boiling point.

Hazardous Materials

1. At what point in the explosive range of flammable gases are explosions of the greatest violence (highest explosion pressures) generally produced?

    (A) at or near the top of the range
    (B) at some intermediate point in the range
    (C) at or near the bottom of the range

2. Which of the following is the most reactive?

    (A) liquid nitrogen
    (B) liquid hydrogen
    (C) liquid fluorine
    (D) liquid oxygen

3. In comparison with the fire hazards of other flammable gases, what is a relatively unique hazard of LPG?

    (A) high density
    (B) low flash point
    (C) high vapor pressure
    (D) high specific heat

4. CHEMTREC is associated with a

    (A) list of flammable liquids.
    (B) list of dangerous chemicals.
    (C) resource center.
    (D) manufacturing plant.

5. In comparison with the density of air, the density of carbon monoxide is

    (A) considerably less; the specific density is .37.
    (B) about half as much; the specific density is .51.
    (C) about the same; the specific density is .97.
    (D) considerably heavier; the specific density is 2.56.

6. The most likely result of using water on burning sodium would be

    (A) that the fire would be extinguished.
    (B) that the sodium would be dissolved.
    (C) an explosive reaction would occur.
    (D) that there would be no change in the intensity of
         the fire.

7.

    From the above diagram it could be concluded that the material in the container would be

    (A) extremely toxic.

(B) extremely flammable.
(C) readily capable of detonation.
(D) radioactive.

8. An orange-colored Department of Transportation hazardous materials warning label indicates that the material shipped is

(A) an explosive.
(B) a flammable liquid or gas.
(C) a poisonous gas.
(D) an oxidizer.

9. While fighting a fire in a large chemical warehouse, you noticed the printed label, "METHANOL" on stacked cartons. All fire officers should realize that this is actually

(A) ethyl alcohol or grain alcohol.
(B) methyl acetate with amyl alcohol.
(C) methyl alcohol or wood alcohol.
(D) methyl chloride.

10. The quantity-distance table known as the "American Table of Distances" is concerned with

(A) ratio of heat or combustion to travel of propagated flame.
(B) location of fire wall in relation to area of hazard.
(C) quantity storage of flammable liquids or gases and distance between storage units.
(D) explosion damage and distance from point of origin.

11.

What does the "W" in the above diagram mean?

(A) The use of water may cause frothing.
(B) Water should be used as the primary extinguishing agent.
(C) Water should only be used in spray or fog form.
(D) Do not use water.

12. Using the Hazardous Identification diagram shown in Question 7 for reference, the space used to identify the health hazard of a material would be what color?

(A) red
(B) blue
(C) yellow
(D) green

13. In order to prevent injury to firefighters, leaking chlorine should be handled by

(A) keeping to the leeward side of the leak.

    (B)   using water freely on the escaping liquid.
    (C)   not throwing loose earth on the liquid.
    (D)   doing none of the foregoing.

14.   "A vapor which is heavier than air, tends to settle in low spots, flows in a stream along a floor and, if reaching a source of ignition, will propagate the flame back to the source for a considerable distance." This description is most appropriate for which of the following?

    (A)   carbon bisulfide
    (B)   illuminating gas
    (C)   acetylene
    (D)   methane

15.   The word BLEVE is most closely associated with

    (A)   a flammable liquid explosion.
    (B)   the rupture of a major container.
    (C)   a cryogenic incident.
    (D)   the release of a toxic substance.

16.   When companies respond to an alarm for a building where fumigation with hydrocyanic acid gas is being conducted, it is imperative that thorough and complete ventilation be effected immediately, primarily because in the concentration usually used

    (A)   rapid corrosion of protective devices occurs.
    (B)   application of water or foam is of limited effectiveness.
    (C)   the gas is explosive even at room temperature.
    (D)   the gas is absorbed through unprotected skin.

17.   Natural gas consists principally of

    (A)   propane.
    (B)   ethane.
    (C)   methane.
    (D)   butane.

18.   Which of the following gases, when found at smoky fires, usually has the most easily noticeable odor?

    (A)   nitrous fumes
    (B)   carbon monoxide
    (C)   carbon dioxide
    (D)   hydrogen sulfide

19.   Which of the following gases is lighter than air and therefore requires ceiling ventilation?

    (A)   benzol
    (B)   ammonia
    (C)   hydrogen sulfide
    (D)   butane

20. A number "2" at the bottom of a Department of Transportation shipping placard without any name means that the material in the container is

   (A) a gas.
   (B) a flammable liquid.
   (C) an explosive.
   (D) an oxidizer.

21. Sulfur dioxide is a toxic gas usually encountered in fires in refrigeration systems, sulfite paper mills, chemical plants, etc. It is detectable when in concentrations of as little as 0.05% because it

   (A) is always accompanied by a light blue smoke.
   (B) is acutely pungent and produces a sense of suffocation.
   (C) is irritating to the hands and face.
   (D) has an odor like that of burning celluloid.

22. While working at a fire in a commercial building, it is learned that a large vat in the area contains 10% acetic acid. The situation should cause you to

   (A) open the vat drain.
   (B) have the firefighters working at the fire use self-contained breathing apparatus.
   (C) operate as though the condition did not exist.
   (D) use as little water as possible to avoid increasing the hazard.

23. The occurrence of unusually high percentages of carbon monoxide at smoldering fires such as furniture factories would probably be primarily due to

   (A) unusually high temperatures.
   (B) the similarity of materials in most cases.
   (C) an oxygen deficiency.
   (D) unusually low temperatures.

24. "Mineral oils at ordinary temperatures are less hazardous than vegetable oils." The best justification for this statement is that, as compared with vegetable oils, mineral oils generally

   (A) have a higher density.
   (B) are more volatile.
   (C) have a lower specific gravity.
   (D) oxidize less readily.

25. Some chemicals do not themselves burn, but when heated liberate oxygen. When involved in fire, such oxidizing materials promote vigorous burning. An example of this chemical is

   (A) acetic acid.
   (B) carbolic acid.
   (C) carbon disulfide

(D)   sodium nitrate.

26. As compared with ordinary commercial gasoline, a fire involving a
     tank of 100 octane gasoline would be

     (A)   very much more hazardous.
     (B)   definitely, if not excessively, more hazardous.
     (C)   of nearly equal hazard.
     (D)   definitely, but not exceptionally, less hazardous.

27. Which of the following is generally considered the most dangerous?

     (A)   gasoline
     (B)   alcohol
     (C)   ethyl ether
     (D)   carbon disulfide

28. Approximately what are the explosive limits of most common flammable
     liquids at room temperature, in terms of percent by volume?

     (A)   0% to 3%
     (B)   1% to 10%
     (C)   15% to 25%
     (D)   30% to 45%

29. Which of the following is a gas that will not burn but is a strong
     supporter of combustion? Oils may ignite explosively at ordinary
     temperatures in the presence of this gas when compressed, and
     density is just slightly more than air, being 1.1.

     (A)   acetylene
     (B)   oxygen
     (C)   carbon dioxide
     (D)   chlorine

30. Cryogenic gases are those gases that

     (A)   react violently with water.
     (B)   are liquefied by pressure.
     (C)   are liquefied by cold.
     (D)   react violently with carbon dioxide.

Answers

1. (B) Explosive range is the range of vapor-air mixture in which burning of the vapor will take place when subjected to a source of ignition. As an example, gasoline has an explosive range of approximately 1.4 to 7.6. This means that only those vapor-air concentrations between 1.4 and 7.6 will burn. The 1.4 means 1.4% vapor and 98.6% air. Vapor-air mixtures below 1.4 are said to be too lean to burn while those above 7.6 are said to be too rich to burn. The flash point of a flammable liquid takes place at the lower explosive limit. At this point, the mixture will flash and go out if subjected to a source of ignition. The highest pressures from the burning mixtures will occur somewhere near the center of the explosive range, or with gasoline with about a 4.7% concentration.

2. (C) Fluorine is the most reactive of the nonmetallic elements. It is corrosive, poisonous, and will cause severe irritation burns to the eyes and skin. It reacts vigorously with almost any substance. Fluorine will even react with concrete and some steels. It must be shipped in specially built containers of steel that contain no impurities.

3. (A) One of the unique hazards of the LPGs is their high vapor densities. When leaks occur in containers of LPG, the vapors tumble to the ground and will travel long distances to a source of ignition. A cloud is usually found around the leak. A good point to be kept in mind is that the vapors within the cloud are too rich to burn. It is at the edge of the cloud that the vapors are within the explosive range. These mixtures are invisible. Consequently, vapor-air mixtures within the explosive limits may travel some distance from the source of the leak and visually not be detected.

4. (C) CHEMTREC stands for Chemical Transportation Emergency Center. It is owned by the Chemical Manufacturers' Association and operates as a public service. CHEMTREC operates around the clock and provides immediate advice for those at the scene of emergencies, then promptly contacts the shipper of the hazardous material involved for more detailed assistance and appropriate follow-up.

5. (C) The question is comparing vapor densities. Air is used as a standard and considered to be 1.0. Although there are slight differences in the vapor density of carbon monoxide, according to different authorities, the vapor density is approximately 1.0, or near that of air. This means that concentrations of carbon monoxide will be found at floor level as well as at ceiling level in some circumstances.

6. (C) If water is used on burning sodium, hydrogen will be given off along with a large amount of heat. The resultant heat will be sufficient to ignite the hydrogen, causing an explosive reaction. It goes without saying that water should not be used on a sodium fire.

7. (C) The numbers used in this identification system are 1, 2, 3, and 4, with 1 indicating the lowest hazard and 4 the highest. The box to the left indicates the health hazard of the material. The 1 indicates only a slight hazard to health. The space at the top

identifies the flammability of the material. The 1 means the material must be preheated before ignition can occur. The space to the right identifies the reactivity of the material. The 4 indicates that the material is readily capable of detonation.

8. (A) A flammable liquid or gas would have a red label.
A poison gas would have a white label.
An oxidizer would have a yellow label.

9. (C) Methanol is methyl alcohol, or wood alcohol. Ethanol is ethyl alcohol or grain alcohol. Ethyl alcohol is used in liquors, but methyl alcohol is poisonous, with death possible from drinking only a small amount.

10. (D) The American Table of Distances is a table used for establishing safeguards for the storage of explosives. In fact, the true name of the table is the American Table of Distances for Storage of Explosives. The table establishes distances that should be maintained between explosive storage and buildings, highways, etc., depending upon the amount of explosives stored.

11. (D) The diagram is one used in the NFPA Hazard Identification System. W̶ (a "W" with a line drawn through it) in the bottom box is used to indicate that water should not be used in the extinguishment of the material.

12. (B) The health indicator would be blue.
The flammability indicator would be red.
The reactivity indicator would be yellow.
The bottom space would be colorless.

13. (B) Chlorine is a nonflammable gas; however, it will react corrosively with many organic materials. It is toxic, having been used as a poisonous gas during warfare. It will also burn the skin if contact is made. Leaking chlorine can best be controlled by water spray. Full protective clothing, including breathing apparatus, should be worn when controlling leaks.

14. (A) Illuminating gas, acetylene, and methane are all gases having a vapor density less than 1.0. These gases will travel toward the ceiling when released. Carbon bisulphide has a vapor density of 2.6. The vapors, therefore, will collect near the floor and travel some distance to find a source of ignition.

15. (B) The word BLEVE is an acronym for "Boiling Liquid Expanding Explosion." As originally set forth by the Factory Mutual engineers it means "a major container failure, into two or more pieces, at a moment in time when the contained liquid is at a temperature well above its boiling point at normal atmospheric pressure." BLEVES are physical occurrences as opposed to chemical reactions and, contrary to what many firefighters believe, the material involved does not have to be flammable, and the BLEVE is not necessarily accompanied by fire.

16. (D) Hydrocyanic acid gas is an extremely poisonous gas that can readily be absorbed through the skin. One good breath would kill

a firefighter. Consequently, when responding to alarms where hydrocyanic acid gas is being used as a fumigant, it is imperative that the building be thoroughly and completely ventilated before any firefighter is allowed to enter. No breathing apparatus can give adequate respiratory protection against this gas.

17. (C) Natural gas is the result of the decomposition of organic material by heat, pressure, and bacteriological action in the absence of air. This usually takes place underground. Natural gas is primarily methane. It is nontoxic and odorless, making it necessary to add an odorant in order to detect a leak.

18. (D) Hydrogen sulfide is the product of incomplete combustion of organic materials containing sulfur. Some of the materials that give off this gas when burning are wool, rubber, and hair. It is usually readily identified by its "rotten egg" smell. Hydrogen sulfide is a poison that attacks the nervous system.

19. (B) Whether or not a gas is lighter or heavier than air can be determined by its vapor density. Gases having a vapor density greater than 1.0 are heavier than air. Those having a vapor density less than 1.0 are lighter than air. The vapor densities of the materials given are:

benzol	2.8
ammonia	0.6
hydrogen sulfide	1.2
butane	2.0

20. (A) The number 2 is a United Nations Number. Other numbers are:

Class No.	UN Class Name
1	Explosives
2	Gases
3	Flammable and combustible liquids
4	Flammable solids
5	Oxidizers and organic peroxides
6	Poisons
7	Radioactive materials
8	Corrosives

These numbers are displayed at the bottom of placards.

21. (B) Firefighters say that sulfur dioxide reaches out and bites you. Although is is a toxic gas, its biting characteristics do not permit a person to voluntarily remain in an area where it is present. If it is necessary, however, to work in a sulfur dioxide atmosphere, breathing apparatus should always be worn.

22. (C) The properties and hazard of acetic acid depend upon the strength of the solution. In diluted form, it is nonhazardous. Consequently, the 10% solution given in the problem is nonhazardous and presents no problem.

23. (C) Carbon monoxide is the product of incomplete combustion. When

sufficient oxygen is present in a fire, the carbon in the burning material will combine with two atoms of oxygen, resulting in the product of complete combustion, which is CO_2 (carbon dioxide). In a smoldering fire where there is an oxygen deficiency, one atom of carbon from the burning material will unite with one atom of oxygen to form CO (carbon monoxide).

24. (D) Vegetable oils oxidize more readily than mineral oils. A good example is linseed oil. The oxidation of linseed oil can lead to spontaneous heating and eventual ignition, if conditions are right.

25. (D) Those chemicals that, when heated, liberate oxygen to promote vigorous burning are called oxidizers. Three of the major classes of oxidizers are the nitrates, nitrites, and chlorates.

26. (C) The octane rating of a gasoline has very little effect upon its hazard. For all practical purposes, gasoline should be considered gasoline regardless of the octane rating. Emergencies involving gasoline with different octane ratings should be treated the same.

27. (D) Carbon disulfide is a material that is in a class by itself. In the UL hazardous classification system, gasoline is rated at 95-100, ethyl ether at 100, kerosene at 40, and carbon disulfide at 110+. This is a material with an extremely low flash point (122), an exceptionally low ignition temperature (212), a wide explosive range (1.3-44), and a high vapor density (2.3). In addition, it is poisonous if swallowed or inhaled, or in prolonged contact with the skin.

28. (B) The explosive limits of most common flammable liquids at room temperature, in terms of percentage by volume, are approximately 1% to 10%. As an example, following are the explosive limits of some common liquids:

Gasoline	1.4-7.6
Jet fuels	1.3-8.0
Kerosene	0.7-5.0

29. (B) The question very clearly lists the characteristics of oxygen. Air is composed of approximately 21% oxygen, 78% nitrogen, and 1% other gases. Oxygen is particularly hazardous when it is liquefied. A leaking cylinder of liquid oxygen would cause an oxygen-enriched atmosphere. Such an atmosphere would add to the intensity of any burning fire.

30. (C) Cryogenics is the liquification of gases by freezing. Two good examples of cryogenic gases are liquid oxygen and liquid nitrogen. Although converting gases to this state saves storage space, it presents additional problems to the fire service. As mentioned in the answer to Question 29, liquid oxygen will greatly accelerate and intensify any fire with which it comes in contact.

Arson

Many states have either adopted the Model Arson Law or enacted legislation similar to this recommended law. However, there are variances among the state laws in penalties and other details regarding arson. In order to remain consistent, answers to questions in this section regarding arson laws are based upon the details outlined in the Model Arson Law.

1. The Model Arson Law was prepared by the

(A) National Fire Protection Association.
(B) International Association of Arson Investigators.
(C) International Association of Fire Chiefs.
(D) Fire Marshals' Association of North America.

2. How many degrees of arson are there in the Model Arson Law?

(A) two
(B) three
(C) four
(D) five

3. How is arson as a crime listed by the Federal Bureau of Investigation?

(A) Class I offense
(B) Class II offense
(C) Class III offense
(D) The Federal Bureau of Investigation does not list
        arson as a crime.

4. Fire officers in charge at a fire must exercise substantial judgment in applying general procedure to specific cases because

(A) of the high property values involved.
(B) no two firefighters proceed simultaneously on a given
        task in the same manner.
(C) all aspects of fire combat require specialization.
(D) no two fires present exactly the same conditioning.

5. The penalty for arson may be more severe in which of the following cases?

(A) The building is occupied by human beings.
(B) The burning occurs at night.
(C) The building is a dwelling.
(D) Human life is endangered.

6. Pyrolysis is primarily concerned with the

(A) intentional setting of fire to a fuel.
(B) chemical change brought about by the action of heat.
(C) rapidity with which a material burns.
(D) products of combustion.

7. If a badly burned body is recovered from a fire, which one of the

following would you <u>not</u> submit as evidence having any potential value in attempting to prove that the deceased individual was alive at the time the fire started?

(A) the presence of soot in the windpipe and lungs
(B) the presence on the skin of liquid-filled blisters, each with a thin red zone around the edge of the burn
(C) the presence of carbon monoxide in the blood
(D) a "pugilistic attitude" of the body; that is, legs and arms partially drawn up as in a fighting position

8. Camara v. Municipal Court of the City and County of San Francisco had a profound effect upon

(A) examining witnesses.
(B) gathering evidence.
(C) obtaining search warrants.
(D) making arrests.

9. An accomplice is

(A) one who, after full knowledge that a felony has been committed, conceals same from the officers.
(B) one who is liable to prosecution for the same identical offense charged against the defendant on trial.
(C) one who harbors a person charged with or convicted of a felony.
(D) a person who has knowledge of a given act.

10. The actual number of arson arrests in comparison to other crimes shows

(A) a relatively low incidence.
(B) a relatively high incidence.
(C) about the same incidence as criminal homicide.
(D) a higher incidence than rape.

11. Arson under the common law is concerned with the willful and malicious burning of

(A) the house of another.
(B) any house.
(C) any building of another.
(D) any building.

12. The person who sets fires for the purpose of stimulating an orgasm or at least for a substitutive sexual act may feel an equivalent thrill from watching the flames. This statement is

(A) commonly believed.
(B) entirely false.
(C) uncertain.
(D) ridiculous.

13. A fire set by a female would most likely be for the purpose of

   (A) attempting to collect insurance.
   (B) revenge.
   (C) sexual gratification.
   (D) covering up another crime.

14. Probably one of the most frequent and most definite clues to arson or attempted arson is

   (A) multiple fires.
   (B) intensity of the flames.
   (C) odor.
   (D) difficulty of extinguishing.

15. Of the following, an arson investigator should first be concerned with the

   (A) origin of the fire.
   (B) latter stages of the fire.
   (C) path of fire travel.
   (D) damage done by the fire.

16. In order for there to be burning of a building there must be a minimum of

   (A) scorching.
   (B) discoloration by heat.
   (C) a single fiber destroyed by fire.
   (D) a considerable amount of charring.

17. In a situation under investigation, honest and conscientious persons

   (A) can be relied upon to give consistently accurate information.
   (B) never agree, even though their sincerity may be proved.
   (C) may give conflicting information.
   (D) will give no information.

18. One kind of arson fire is that set to cover up another crime. The most common crime attempted to be covered up by a fire is

   (A) murder.
   (B) embezzlement.
   (C) tax fraud.
   (D) burglary.

19. Conviction for the crime of arson in connection with which one of the following structures would carry the most severe penalty?

   (A) private residence
   (B) theater
   (C) school
   (D) church

20. A person confesses setting an incendiary fire. When is the confession considered enough to establish the criminal origin of the fire?

    (A) always
    (B) usually
    (C) sometimes
    (D) never

21. In order to establish the corpus delicti of arson, it must be established that

    (A) there was actual burning.
    (B) the fire was maliciously set.
    (C) the fire was willfully set.
    (D) all three of the above elements existed.

22. In the collection and preparation of evidence to be presented in court, the investigator should remember that the presumption of the court concerning the origin of the fire is that

    (A) a criminal agency was responsible for the cause of the burning.
    (B) when a building is burned it is the result of an accident.
    (C) a confession made out of court by the accused is sufficient proof of the motive.
    (D) a building which was burned has been willfully fired by some responsible person.

23. In two separate fires, one set by a pyromaniac and the other set by the owner of a business which was proving to be a losing venture, the primary distinguishing feature between the two fires would be the

    (A) method by which each of the two fires was started.
    (B) device used as the source of ignition in each fire.
    (C) fire in the business place was of accidental origin, whereas the pyromaniac's fire was intentional in origin.
    (D) motivation factor in each of the fires.

24. It can best be said that the corpus delicti of arson is proof

    (A) that the fire was set.
    (B) that there was actual burning.
    (C) connecting the suspect with the burning.
    (D) of the incendiary origin of the fire.

25. The Escobedo and Miranda decisions are concerned with

    (A) search warrants.
    (B) witnesses.
    (C) interrogation of a suspect.
    (D) the corpus delicti of arson.

26. Ammonia is generally used by an arsonist to

(A) cause an explosion in a building.
(B) start a fire in several different places at the
same time.
(C) eradicate fingerprints and footprints.
(D) render unnoticeable the odor of gasoline.

27. Perhaps no other motive is so frequently responsible for the commission of arson as

(A) revenge.
(B) spite.
(C) jealousy.
(D) profit.

28. The statute of limitations as concerned with criminal cases refers to the

(A) limitations of being tried twice for the same
offense.
(B) extent to which a judge may go in imposing a sentence.
(C) admissibility of certain testimony as evidence.
(D) duration of time in which a prosecution for an
offense may be commenced.

29. Research has shown that about the minimum time required for a cigarette to cause flaming combustion when dropped into an overstuffed chair or sofa is

(A) one-half hour.
(B) one hour.
(C) one and one-half hours.
(D) two hours.

30. During the investigation of a fire, the arson investigator reaches the conclusion that A set fire to B's house to get even with B for imagined wrongs. However, try as he would, the investigator could not obtain concrete facts or evidence to support his theory. When writing his report, the best thing for the investigator to do is to

(A) set forth his beliefs in detail in the report.
(B) make no report and consider the case closed.
(C) set out the facts in the report, and outline his
theories in a separate letter.
(D) wait a few weeks to see if the case will break before
doing anything about it.

31. The physical characteristics of the grudge fire most nearly resembles that of the

(A) vanity fire.
(B) insurance fire.
(C) pyromaniac fire.
(D) fire set to cover up another crime.

32. A fire in which flammable liquids have been thrown under a porch would most likely have the motivating factor of

   (A) revenge.
   (B) covering up another crime.
   (C) insurance.
   (D) pyromania.

33. When an arson investigator has completed an investigation, the best procedure for the investigator to follow would be to

   (A) write down all the important facts of the investigation before leaving the scene.
   (B) fix in mind all important facts of the investigation so that he or she can make out the report at the end of the day.
   (C) return immediately to the office and write a report of the investigation from notes and memory.
   (D) require each of the persons interviewed in the course of the investigation to make a brief written report of the information he or she has contributed.

34. An arson investigator, when checking a sofa that had burned during a fire, noticed that the springs were collapsed. This should give the investigator reason to believe that

   (A) an accelerant was used in the fire.
   (B) the fire burned very quickly.
   (C) the fire was of a smoldering nature.
   (D) the fire was not caused by a cigarette.

35. In trying to establish the fact of the incendiary origin of a fire by direct evidence, which, if any of the following, would you consider as not pertinent to that purpose?

   (A) the arrangement and position of stock and material
   (B) the arrangement of the furniture
   (C) the number of separate and distinct fires on the premises at the same time
   (D) the size of the insurance policy

36. The first step in establishing the cause of a fire is to

   (A) locate the point of origin.
   (B) determine if forcible entry was made.
   (C) check for the use of an accelerant.
   (D) question witnesses.

37. Circumstantial evidence in the proving of arson is

   (A) admissible
   (B) not admissible.
   (C) not valuable.
   (D) never used.

38. Tests have shown that most woods have a charring rate of approximately

    (A) one-half inch per hour.
    (B) one inch per hour.
    (C) one and one-half inches per hour.
    (D) two inches per hour.

39. Arson is a crime in which it is difficult to secure evidence, because the

    (A) evidence can readily be removed.
    (B) evidence is usually consumed.
    (C) crime cannot be reconstructed.
    (D) investigator does not know what to look for.

40. Of the various motives for setting fires, the one which accounts for the largest percentage of loss from incendiarism is

    (A) grudge fires.
    (B) vanity fires.
    (C) insurance fires.
    (D) pyromaniac fires.

Answers

1. (A) The Model Arson Law was prepared by the National Fire Protection Association in 1931. Some states have adopted this recommended
law almost verbatim while others have used it as a guide in the en-
actment of their own arson laws.

2. (C) There are four degrees of arson in the Model Arson Law. Arson
--First Degree refers to the burning of dwellings. Arson--Second
Degree is concerned with the burning of buildings other than dwell-
ings. Arson--Third Degree involves the burning of other property.
Arson--Fourth Degree is concerned with attempts to burn buildings
or other property.

3. (A) For many years, the Federal Bureau of Investigation did not
list arson as a major crime. Arson has been re-evaluated and is
now listed as a Class I offense.

4. (D) Although all fires are somewhat similar, no two fires are
exactly alike. Consequently, it is difficult to generalize about
fires or to apply any general procedures to specific cases. Each
fire must be fought and investigated based upon the particular factors presented.

5. (C) The arson laws do not address whether or not a building is occupied or human life is endangered; neither do they distinguish as
to whether a fire is set at night or in the daytime. The penalty
established is based entirely upon the kind of building burned. The
penalty for setting fire to a dwelling is more severe than for setting fire to other kinds of buildings.

6. (B) Pyrolysis is the chemical change brought about by the action
of heat. It is important that arson investigators be familiar with
the pyrolysis of the various fuels encountered during investigations
in order to be completely knowledgeable about the burning process
and burn patterns.

7. (D) Answers (A), (B), and (C) indicate that the person was alive
at the time of the fire. The attitude of the body has nothing to
do with whether the person was alive or dead when the fire burned
it. Incidentally, soot in the windpipe and lungs indicate the person was alive at the time of the fire. If soot is in the stomach,
it is an indication that the person was probably dead at the time
of the fire. If no carbon monoxide is found in the blood, it is a
fairly positive sign that the person was dead at the time of the
fire.

8. (C) This case and See v. City of Seattle had a profound effect upon
obtaining search warrants. These cases established that warrants
are necessary in order to make fire inspections and fire investigations under certain circumstances. The Camara case had to do with
residences while the See case had to do with a commercial occupancy.
It is important that those responsible for fire prevention inspections and fire investigations be aware of the decisions of these
two cases and how the decisions and opinions affect department operations.

9. (B) An accomplice is very well defined in the arson laws. As an example, in Arson--First Degree it says, "Any person who willfully and maliciously sets fire to or burns or causes to be burned or who aids, counsels, or procures the burning of . . . shall be guilty of Arson in the first degree." As shown in this section, an accomplice is just as guilty as if he or she set the fire.

10. (A) Although arson has been estimated to be the number one crime in the United States from a dollar loss standpoint, the actual arrests made when compared with the number of crimes committed is relatively small. With the effort to combat arson on the increase, the number of arrests made will no doubt increase; however, due to the nature of the crime the number of arrests will probably remain relatively low.

11. (A) The definition of arson under the common law is very restrictive. It refers only to the burning of the house of another person. Under the common law, a person could burn his own house and not be guilty of arson.

12. (A) Although many people believe that pyromaniacs obtain a thrill from watching the flames, it has been found that these people obtain their satisfaction from applying the match and watching it only long enough to make sure the fire will burn. After that, their attention centers on the firefighters and the action that ensues.

13. (B) The proportion of female fire setters as compared with males is very small. Females generally set fires for revenge or to attract attention. They seldom set fires outside their own personal sphere of interest. This sphere of interest normally includes the home, the residence of a lover, a church, their place of work, or a neighbor's house.

14. (A) In almost all cases where two or more separate fires are found, it readily establishes the corpus delicti. In fact, some courts have held that simultaneously burning fires are prima facie evidence of an incendiary origin. Many arsonists will not rely on only one fire to completely destroy a building, and, therefore, will start two or more with the intent that the fires will burn together and destroy the evidence that multiple fires had been involved. Fortunately, a number of the fires are discovered early, giving the fire department a chance to extinguish the fires before they burn together.

15. (A) Although all the factors listed are important, an arson investigator must first be concerned with the origin of the fire. An investigator who spends time concentrating on the latter stages of the fire is wasting valuable time because little can be done to determine the cause and fire travel until the origin of the fire is determined.

16. (C) The courts have held that a building has not been burned if there is merely scorching; the building is smoked up; or the building has been discolored by heat. However, if as much as a single fiber of wood is destroyed by fire, it is sufficient to establish that the building has been burned.

17. (C) Honest and conscientious persons may give conflicting information; however, the conflicting information is usually not intentional. Two people seeing the same accident will give different information as to what they saw and each will be sure that what he or she reports is factual. A basic truth about human beings is that the perception of each individual is not the same.

18. (D) Fires have been set in an attempt to hide a murder, embezzlement, tax fraud, inventory shortages, vandalism, and other crimes. However, the most common kind of crime fire attempts to cover is burglary.

19. (A) Burning of a private residence is a violation of Arson--First Degree which calls for a penalty of not less than two or more than twenty years. Burning of a theater, school, or church is a violation of Arson--Second Degree that is punishable by not less than one or more than ten years.

20. (D) The courts have held that a confession by itself is not sufficient to establish the corpus delicti (body of the crime). However, if the corpus delicti has been established, the confession can be used in connection with other evidence in order to obtain a conviction.

21. (D) To establish the corpus delicti of arson, it must be shown that burning actually occurred, and that the fire was willfully and maliciously set. If any one of these three elements is missing, then there is no corpus delicti.

22. (B) The courts have related the cause of fires to the presumption of the innocence of an accused. Rather than presuming that a fire has been intentionally set, the courts have held that a fire carries with it the presumption that it is of accidental or providential origin. Consequently, in many arson cases it is necessary to eliminate all nature causes in order to prove that a fire was intentionally set.

23. (D) Both fires would be arson; however, the primary distinguishing feature between the two would be the motivating factor. While the owner of a business would set fire to the business to collect the insurance, the pyromanic would set fire to a building because of the impulse to apply a match to a flammable object.

24. (D) The corpus delicti of arson is proof of the incendiary origin of the fire. The corpus delicti is the body of the crime. The elements of the corpus delicti of arson is that property has been burned, and that the burning was done willfully and maliciously.

25. (C) These decisions involve the rights of a suspect in an interrogation. They require that a suspect must be given the following warnings:

1. You have the right to remain silent.
2. If you give up that right, anything you say can and will be used against you in a court of law.
3. You have a right to talk to a lawyer and have him

present with you while you are being questioned.
4. If you cannot afford to hire a lawyer, one will be appointed for you, without any cost to you, prior to questioning.

26. (D) In addition to using ammonia, some arsonists use perfume and similar strong odorants to try to hide the smell of an accelerant. These liquids are sometimes effective in concealing the odor of an accelerant, but their detection in areas where they should not be is a cause to suspect that the fire was due to arson.

27. (D) It is generally considered that "selling out to the insurance company" is the number one motivating factor for arson fires. The number of insurance fires is on the increase with many of the larger fires being set by professional arson rings. Insurance fires increase as business conditions worsen.

28. (D) A statute of limitations is concerned with the duration of time in which a prosecution for an offense may be commenced. As an example, if the statute of limitations is five years, then prosecution of the person who committed the crime must be started within five years or the person is free and clear of the crime, regardless of the guilt.

29. (C) Cigarettes are constructed so that they will continue to burn after being discarded. When one is dropped into an overstuffed chair or sofa, it will bury itself and cause a heat buildup. Heat will continue to develop until a sufficient amount has built up to cause flame propagation. This normally takes one and one-half to two and one-half hours.

30. (C) Fire reports should contain only facts. Only facts can be submitted to court. However, it is good practice for an investigator to outline his or her theories in a separate report and file the report for future reference. It is difficult to recall a thought process at a later date, and the information outlined in the separate report might be beneficial in the future.

31. (C) The physical characteristics of the grudge fire most nearly resemble those of the pyromaniac fire. In many cases, kerosene is used in both kinds of fire. Although the grudge fire is a single attack against one person, the pyromaniac operation is a repetition of fires.

32. (A) This would most likely be a revenge fire. Those trying to get even by damaging a dwelling will sometimes use flammable liquids under porches, in crawl spaces under houses, or in garages. If flammable liquids are not used, they will sometimes stuff paper under the porch or in the crawl spaces.

33. (A) It is easy to forget important facts if they are not written down as soon as possible. Additionally, arson investigators might be dispatched to another alarm before returning to the office. Not only would the investigator probably forget some of the facts from the first fire, but he or she might also intermingle the details on the two fires. Consequently, it is always best to write down

all important facts before leaving the scene and also to obtain
photographs of all pertinent situations.

34. (C) The condition of the springs is an indication of whether the
fire burned quickly or slowly. The springs of an overstuffed chair
or sofa are likely to flatten out if the fire was slow burning.
This is usually the case if a cigarette is accidentally dropped
into a chair or sofa. If an accelerant is used, the fire will burn
quickly and the springs will not become anneled; consequently, they
will maintain their shape.

35. (D) The size of the insurance policy is not evidence. It may be
used to help establish a motive for the fire, but it is not related
to the establishment of the corpus delicti. However, the arrangement of furniture to assist in maximizing the fire, or multiple
fires on the premises can be used as direct evidence to help establish the incendiary origin of the fire. Multiple fires are usually
considered prima facie evidence of the incendiary nature of the fire.

36. (A) The first step in establishing the cause of a fire is to locate the point of origin. Talking to witnesses, checking for the
use of an accelerant, and determining if forcible entry was made
are all important to the investigation, but the point of origin
must first be established when attempting to establish the fire
cause.

37. (A) Direct evidence is eyewitness evidence or a confession. Seldom in arson cases will direct evidence be used to obtain a conviction. Arson is a crime that is usually committed in secret without
any witnesses; consequently, it is necessary in the majority of
cases to rely upon circumstantial evidence to establish the guilt
of the accused.

38. (C) Tests conducted by the United States Forest Products Laboratory indicated that most woods have a uniform char development of
about 1.54 inches per hour. These results were obtained when the
wood was exposed to an average flame temperature of from 1400°F to
1600°F.

39. (B) Unfortunately, one of the difficulties of establishing the corpus delicti of arson is that evidence is destroyed by the fire. This
is one of the primary reasons why fewer arrests are made for arson
than for other kinds of crime. Without evidence to establish the
corpus delicti, it cannot be proved that a crime has been committed.

40. (C) The loss from insurance fires by far accounts for the largest
percentage of loss from incendiarism. These fires follow trends.
Insurance fires increase when the economy is down and decrease when
the economy is good. "Selling out to the insurance company" has
reached the stage where professional arson rings are used to ensure
a good burn.

VI

BUILDING CONSTRUCTION AND SYSTEMS

Building Construction

1. The primary purpose of fire-resistive construction of a building is to

   (A) protect the structure so that it is easily made re-
       usable after a fire.
   (B) retard the spread of fire and facilitate escape of
       occupants.
   (C) protect the building from exposure fires.
   (D) reduce the likelihood of fires starting.

2. Most nearly, what is the minimum width of exit usually required for a single file of persons?

   (A) 15 to 17 inches
   (B) 18 to 20 inches
   (C) 21 t0 23 inches
   (D) 24 to 27 inches

3. Of the following structure factors of a building, which one is least important from a fire spread standpoint?

   (A) vertical openings
   (B) foundation
   (C) walls
   (D) horizontal openings

4. When considering the possible results of the collapse of a non-load bearing ceiling during a fire, which of the following can best be said?

   (A) The walls will come down with the ceiling.
   (B) The roof, if the ceiling is part of the top floor,
       will come down with the ceiling.
   (C) Both the walls and the roof are likely to come down
       with the ceiling.
   (D) The ceiling will collapse without having a major
       effect on either the walls or the roof.

5. Considering both girders and beams, it can best be said that

    (A) the two terms are the same.
    (B) girders rest on beams.
    (C) beams rest on girders.
    (D) girders and beams run parallel.

6. In lightweight wood construction, how far apart are floor joists normally spaced?

    (A) 12 inches, center-to-center
    (B) 16 inches, center-to-center
    (C) 24 inches, center-to-center
    (D) 30 inches, center-to-center

7. The aide to the officer-in-charge of a fire reports that there has been a collapse of a party wall. The officer-in-charge should immediately know that this wall is

    (A) a non-bearing wall.
    (B) a bearing wall.
    (C) the division point between two buildings.
    (D) located on the exterior of the building.

8. Balloon wood frame construction is referring primarily to the way the

    (A) roof is attached to the top plate.
    (B) kinds of wall studs used.
    (C) ceiling arrangement of the structure.
    (D) manner in which the sill is attached to the foundation.

9. It can normally be expected that a panel wall between two rigid supports which is subjected to a fire will tend to

    (A) bow toward the fire.
    (B) bow away from the fire.
    (C) remain in a rigid condition.
    (D) disintegrate.

10. Most firefighters are aware of the possibility of exterior walls collapsing under extreme fire conditions. They should also be aware that with the collapse of an ordinary masonry exterior wall, the wall will probably fall a distance

    (A) one-third of its height.
    (B) one-half of its height.
    (C) three-fourths of its height.
    (D) equal to its height.

11. Concrete used in building construction is many times subjected to tension, compression, and shear forces. Under normal conditions, it can be said that concrete can best withstand

    (A) tension forces.

(B)   compression forces.
(C)   shear forces.
(D)   tension, compression, and shear forces in equal
        amounts.

12.   Concrete subjected to extreme heat will spall. This means that it
      will

(A)   come apart with explosive force.
(B)   crumble.
(C)   harden.
(D)   run as if melted.

13.   Ordinary construction most nearly refers to the

(A)   material used in the walls.
(B)   size of beams and girders.
(C)   kind of roof used on the building.
(D)   manner in which vertical openings are protected.

14.   There are two primary factors affecting the stability of unprotec-
      ted steel beams when subjected to fire. One is the elongation of
      the beam and the other is the collapse of the beam. It should be
      expected that elongation of the beam will occur when the beam is
      subjected to temperatures of

(A) 900°F to $1,100^\circ$F.
(B) $1,100^\circ$F to $1,300^\circ$F.
(C) $1,300^\circ$F to $1,500^\circ$F.
(D) $1,500^\circ$F to $1,600^\circ$F.

15.   It should be expected that a lintel in a building will be found

(A)   on the roof.
(B)   in the basement.
(C)   over a window opening.
(D)   between studs in a wall for the purpose of fire stop-
        ping.

16.   One of the characteristics of mill construction is that

(A)   it is slow burning.
(B)   it is fast burning.
(C)   all columns are made of concrete.
(D)   all beams are made of concrete.

17.   Scuppers are designed into buildings for the purpose of

(A)   early warning of a fire.
(B)   water removal.
(C)   aiding the torsional resistance.
(D)   smoke removal.

18.   The NFPA rates roof coverings as Class A, Class B, etc. A Class A
      roof covering would

(A) have a two-hour fire-resistive rating.
(B) have a one-hour fire-resistive rating.
(C) have little, if any, resistance against fire.
(D) be effective against severe fire exposures.

19. The fire load of a building is generally expressed in terms of the

(A) weight of combustible material per 100 square feet of fire area.
(B) weight of combustible material per square foot of fire area.
(C) total weight of combustible material within a given fire area.
(D) total weight of combustible material within a structure.

20. The Steiner Tunnel Test is used to determine the

(A) strength of structural steel.
(B) rating of roofing materials.
(C) flame spread of materials.
(D) strength of masonry materials.

Answers

1. (B)   The primary purpose of making buildings fire-resistive is to retard fire spread and facilitate the escape of occupants.   That is why so much emphasis is placed on the protection of vertical openings, protection of horizontal openings, limitation of fire area, and exits.

2. (C)   This same principle is discussed in the area of fire prevention; however, it is worth duplicating.   The NFPA standards regarding exits provides for an exit width of 22 inches for each single file of persons from buildings.   This standard is used in determining exit requirements in relationship to the expedient and safe removal of occupants from buildings.

3. (B)   Although it is extremely important to the overall structural safety of a building, the foundation has little effect upon the fire spread within a building.

4. (D)   A non-load bearing ceiling supports only its own weight.   A collapse of the ceiling should have little, if any, effect on the walls or roof.

5. (C)   Girders and beams run in opposite directions with the girders supporting the beams.   The collapse of a beam would cause a localized problem; however, the collapse of a girder would magnify the problem.

6. (B)   This is generally the requirement in most building codes.   It should be remembered that wall studs are also normally spaced 16 inches, center-to-center.

7. (C)   A party wall separates two buildings.   The two buildings may or may not belong to the same owner.   Party walls can be either nonbearing or bearing walls.

8. (B)   In some wood construction, multistory buildings, the wall studs only extend from floor to floor.   With balloon construction, the wall studs extend continuously for two or three floors.

9. (A)   A wall panel, when subjected to heat, will tend to expand on the side exposed to the fire.   Consequently, a panel will tend to bow towards the fire.

10. (A)   In addition to the collapsed wall falling within a distance equal to one-third of its height, it can be expected that bricks or pieces of the masonry may bounce and roll much farther.

11. (B)   Tension refers to the ability of a material to withstand being pulled apart.   Compression refers to the ability of a material to withstand being compressed.   Shear is the ability of a material to withstand being broken apart by a side force.   Concrete is best capable of withstanding compression forces.

12. (B)   Spalling in concrete is caused by the expansion and subsequent drying-out of moisture.   This causes a crumbling effect with the

possibility of early collapse of a structure. In some cases, spalling occurs in an explosive form.

13. (A) There are several factors related to ordinary construction; however, the prime factor is that the walls are made of masonry.

14. (A) Elongation of an unprotected steel beam in a fire could contribute to a collapse of the building if the beam is secured at both ends. Elongation will commence somewhere in the neighborhood of 900°F to $1,100^{\circ}$F. In serious fires where the temperature may reach above $1,500^{\circ}$F, total collapse of the beam will probably occur.

15. (C) A lintel is a beam that carries the load of a wall over an opening such as a door opening or a window opening. Lintels may be made of wood, metal, or masonry.

16. (A) Another term for mill construction is heavy timber construction. One characteristic of this construction is slow burning. The columns and beams in mill construction are extremely heavy.

17. (B) There are two devices designed into buildings to assist in water removal. These devices are called scuppers and floor drains. Floor drains are much more likely to become clogged than are scuppers. When scuppers are designed into exterior walls, they should not be installed directly over window openings of the lower floors.

18. (C) The NFPA roofing ratings are as follows:

> Class A--Effective against severe fire exposures.
> Class B--Effective against moderate fire exposures.
> Class C--Effective against light fire exposure.

19. (B) Fire load is used to estimate the magnitude of a fire which might occur in a fire area. It is expressed as the weight of combustible material per square foot of the fire area. Both the combustible structural members and the combustible contents are considered when determining the fire load.

20. (C) The Steiner Tunnel Test is used to determine the flame spread characteristics of various materials. The test has been adopted by the NFPA, and in the Life Safety Code materials are classified in accordance with results obtained from the test.

Fire Protection Systems

1. What percentage is, most nearly, the recorded effectiveness of sprinkler systems in the United States?

    (A) 99%
    (B) 95%
    (C) 90%
    (D) 85%

2. One kind of carbon dioxide fire protection system is referred to as a total flooding system. This kind of system must be equipped with

    (A) a predischarge alarm.
    (B) a manual control valve.
    (C) an automatic infrared device.
    (D) at least ten discharge heads.

3. An objection raised against the use of a dry pipe system of automatic sprinklers as compared to a wet pipe system is that in the former system

    (A) it is more difficult to get complete coverage.
    (B) the time between the opening of a sprinkler and the
         issuance of water is greater.
    (C) the water drains off too easily.
    (D) freezing, with possible crippling of the system,
         occurs more frequently.

4. The most common cause of accidental water damage from automatic sprinklers is

    (A) knocking off of a head.
    (B) overheating.
    (C) mechanical failure of a head.
    (D) improper maintenance.

5. A fire department connection would most likely be found in

    (A) a wet pipe sprinkler system.
    (B) a dry pipe sprinkler system.
    (C) a deluge sprinkler system.
    (D) all of the above sprinkler systems.

6. Some dry chemical extinguishing systems are designed for compatibility with foam. The dry chemical used in these systems is basically a

    (A) potassium bicarbonate agent.
    (B) sodium bicarbonate agent.
    (C) potassium chloride agent.
    (D) monoammonium phosphate agent.

7. The best general protection for areas where proxylin or other nitrocellulose materials are stored or handled is provided by

    (A) soda acid extinguishers.

(B)   carbon dioxide.
(C)   foam.
(D)   automatic sprinklers.

8.   A picker trunk automatic sprinkler head is used primarily

(A)   outside buildings.
(B)   in high-temperature ovens.
(C)   inside ducts.
(D)   in occupancies with high ceilings.

9.   Which of the following differentiates dry type from wet type sprink-lers?

(A)   a low water pressure
(B)   a high water pressure
(C)   use of air under pressure
(D)   size

10.   The dry chemical known as purple K has a

(A)   sodium bicarbonate base.
(B)   potassium bicarbonate base.
(C)   monammonium phosphate base.
(D)   potassium chloride base.

11.   A pendent sprinkler head is one installed

(A)   in an upright position.
(B)   in a down position.
(C)   primarily in limited supply sprinkler systems.
(D)   outside a building.

12.   The dry chemical agent Met-L-X has a

(A)   sodium bicarbonate base.
(B)   potassium bicarbonate base.
(C)   monammonium phosphate base.
(D)   sodium chloride base.

13.   When a pumper reported to a fire, it was assigned by the chief-in-charge to hook a line immediately to the sprinkler siamese of the building adjacent to the fire building. Of the following, the most likely reason for this order was to

(A)   prevent a backflow of water from the normal sprinkler
         supply to the water main.
(B)   provide the pumper with an added supply in the event
         that use of the main proves excessive.
(C)   provide more positive control over water flow from
         the sprinklers.
(D)   insure water at sufficient pressure to the sprinklers.

14.   What is the minimum amount of water recommended for a pressure tank on a limited water supply sprinkler system when the system is protecting a light hazard occupancy?

(A) 1,000 gallons
(B) 2,000 gallons
(C) 3,000 gallons
(D) 5,000 gallons

15. Of fires in buildings equipped with automatic sprinklers, what percentage, most nearly, were either held in check or extinguished by the sprinklers?

(A) 25%
(B) 50%
(C) 75%
(D) 95%

16. For full effectiveness, a total flooding dry chemical fire protection system must produce the desired concentration within a reasonable period of time. Within what period of time does the NFPA recommend?

(A) 10 seconds
(B) 15 seconds
(C) 20 seconds
(D) 30 seconds

17. An automatic sprinkler system employing open sprinklers attached to a piping system, connected to a water supply through a valve, which is opened by the operation of a heat responsive system installed in the same areas as the sprinklers, is called a

(A) pre-action system.
(B) wet pipe system.
(C) dry pipe system.
(D) deluge system.

18. A differential valve is used in a

(A) wet pipe sprinkler system.
(B) dry pipe sprinkler system.
(C) deluge sprinkler system.
(D) pre-action sprinkler system.

19. The purpose of a post indicator valve is to

(A) indicate the pressure on the water supply side
       of a sprinkler system.
(B) indicate the pressure on the system side of a
       sprinkler system.
(C) control the flow of water in a sprinkler system.
(D) provide a means for the fire department to pump
       into a sprinkler system.

20. In regard to where the riser feeds the sprinkler system, there are four methods which can be employed: center central feed; side central feed; central end feed; and side feed. In a system in which there are over six sprinklers on a branch line, which method of feeding the system is recommended?

    (A)   center central feed
    (B)   side central feed
    (C)   central end feed
    (D)   side end feed

21.   A number of different kinds of halogenated agents have been used in fire protection systems. Which is the most common of these?

    (A)   Halon 1202
    (B)   Halon 1211
    (C)   Halon 1301
    (D)   Halon 2402

22.   A wet pipe sprinkler system is one that has water in it

    (A)   at all times under pressure.
    (B)   at all times, but pressure is not applied until such
         time as a sprinkler head opens.
    (C)   at all times that may or may not be under pressure.
    (D)   under at least 100-psi pressure.

23.   The high piling of combustible material at sprinkler heads is undesirable principally because

    (A)   sprinklers fail to operate.
    (B)   water cannot reach the seat of the fire.
    (C)   there is more content capable of burning.
    (D)   aisles are obstructed.

24.   How many of the following kinds of sprinkler systems employ heat activating devices: wet pipe; pre-action; deluge; combined dry pipe; and pre-action?

    (A)   one
    (B)   two
    (C)   three
    (D)   four

25.   Sprinkler heads in severely corrosive atmospheres are usually coated with

    (A)   lacquer.
    (B)   shellac.
    (C)   lead.
    (D)   wax.

26.   A gren-gun is a portable fire extinguishing system that fires projectiles filled with

    (A)   carbon dioxide.
    (B)   dry chemical.
    (C)   water.
    (D)   Halon 1301.

27.   What temperature rating would a sprinkler head with the frame arms colored red have?

(A) 165°F
(B) 212°F
(C) 286°F
(D) 360°F

28. "...system is an arrangement of automatic sprinklers attached to a piping system containing air that may or may not be under pressure ..." Which system meets the above description?

(A) wet pipe
(B) pre-action
(C) dry pipe
(D) deluge

29. Which is the most common kind of sprinkler system used throughout the United States?

(A) wet pipe system
(B) dry pipe system
(C) pre-action
(D) deluge

30. Of the following, which is the primary reason sprinkler systems fail to extinguish or control a fire?

(A) sprinkler system shut off
(B) inadequate maintenance
(C) inadequate water supply
(D) building only partially protected

Answers

1. (B) The effectiveness of sprinkler systems is based upon whether or not a fire is extinguished or held in check by the sprinkler system. Based upon information submitted to the NFPA, the overall effectiveness of sprinkler systems in the United States is approximately 95%. However, because of the number of unreported fires in sprinklered buildings where only one to three heads are used to extinguish or control the fire, it is estimated that the effectiveness of sprinkler systems is actually much higher than 95% and may even be above 99%.

2. (A) A total flooding system must be designed to provide for the evacuation of all occupants from the area prior to the discharge of the CO_2. This requires that the system be equipped with a predischarge alarm and a discharge delay that will allow sufficient time for the occupants to leave the area.

3. (B) Water is projected onto the fire or in the fire area immediately when the head ruptures in a wet pipe system. After the head ruptures, it is necessary for the system to fill with water before any water is placed on the fire in a dry pipe system. Consequently, it can be expected that the fire loss will be greater with a dry pipe system due to this time delay.

4. (B) Overheating is a term used to denote an abnormally high temperature at ceiling level without the presence of fire. This condition generally develops as a result of hot manufacturing processes, lack of ventilation, or artificial heating such as heat from steam pipes, weather. Many times, conditions develop due to a change in manufacturing processes without any thought being given to the effect the change will have on ceiling temperatures.

5. (D) A fire department connection is required on all sprinkler systems. Piping from the fire department connection connects into the sprinkler system on the system side of the shutoff valve. This permits the fire department to pump into the system even if the main shutoff valve has been closed.

6. (B) When systems are designed for compatible use with foam, a sodium bicarbonate agent is used. The sodium bicarbonate agent is generally treated with silicone to improve the flow of the powder. The system must be designed to limit the disturbance of the foam blanket when the dry powder is applied.

7. (D) Nitrocellulose materials provide their own oxygen and, consequently, fires in these materials cannot be extinguished by excluding the air. The best extinguishing material for these products is water in large quantities. The quickest and most effective method of supplying water is through the use of an automatic sprinkler system.

8. (C) Picker trunk sprinkler heads are designed to reduce the collection of lint or fibers. They are equipped with a small, smooth deflector. These heads are used inside ducts where foreign material may be found in suspension.

9.   (C)   A wet pipe sprinkler system has water in the system under pressure at all times.   A dry pipe sprinkler system has air in the system under pressure at all times.

10.   (B)   Purple K (PKP) was developed by the United States Naval Research Laboratory.   It has a potassium bicarbonate base.   Purple K is about two and one-half times as effective as a sodium base dry powder on flammable liquid spills.

11.   (B)   A pendent sprinkler head is one installed in a down position. The effectiveness and distribution of a pendent head is almost identical to a head installed in the upright position.

12.   (D)   Met-L-X dry chemical agent has a sodium chloride base.   A thermoplastic additive together with other additives is used to provide for moisture resistance and to develop the essential free-flowing characteristic.

13.   (D)   All sprinkler systems are equipped with fire department inlets for pumping into the system.   These inlets connect into the sprinkler system on the system side of the shutoff valve.   When the fire department pumps into the system, in addition to insuring adequate pressure to the sprinkler heads, it provides water to open heads in the event the system has been shut down.

14.   (B)   This is a NFPA recommendation.   Limited water supply systems are used in areas where conventional supplies such as a public water system or a gravity tank are not available.   Although the minimum recommended amount of water for a light hazard occupancy is 2,000 gallons, the minimum recommended for an ordinary hazard occupancy is 3,000 gallons.

15.   (D)   A record of 95% is extremely good, but the actual efficiency is probably even better.   The 95% record only includes those fires in sprinklered buildings which were reported to the NFPA.   Many fires where only one or two heads are used to control the fire are never reported.   If all fires in sprinklered buildings were reported, the overall efficiency would probably approach 99% or better.

16.   (D)   The NFPA recommends that the system be designed so that the dry chemical is applied at a rate to produce the desired concentration throughout the entire area to be protected within 30 seconds.

17.   (D)   The system described is a deluge system.   The fire is detected by a HAD (heat actuating device) that is located in close proximity to the sprinkler heads.   Detection of the fire by a HAD will cause the main supply valve to open.   The entire system is immediately filled with water with the result that water is projected from all heads on the system, completely flooding the area.   The system is used in occupancies where the chance of a rapid spreading fire exists, such as where flammable liquids are used, or perhaps in a plant handling explosives.

18.   (B)   A differential valve is used in a dry pipe sprinkler system. The differential valve is designed to permit a small amount of air pressure to hold back a larger amount of water pressure.   This

is generally accomplished by having a large diameter air clapper in the valve to bear pressure directly on a smaller water clapper valve. If the pressure on both sides of the valve is equal, the force on the air clapper valve is more than that on the water clapper valve due to the different areas of the two valves.

19. (C) A post indicator valve (PIV) is a sprinkler shutoff valve. It is generally located outside a building, somewhere between the source of water and the building. The PIV valve has a permanently installed wrench that can be used to open the valve, and it is equipped with a window to indicate whether the valve is open or closed.

20. (A) The riser is near the midpoint of all the sprinkler heads when a center central feed is used. This provides a better flow to all heads and permits a better distribution in the event all heads open.

21. (D) All the halogenated agents listed have been used in extinguishing systems. The most common agent used throughout the United States is Halon 1301. Halon 1301 is bromotrifluoromethane. All of the halogenated agents are believed to extinguish a fire by breaking the chain reaction.

22. (A) A wet pipe sprinkler system is one that has water in it under pressure at all times. When a sprinkler head is opened, water is immediately discharged.

23. (B) Piling combustible material too close to sprinkler heads will many times prevent the water from reaching the seat of the fire. It is good practice to restrict the piling of material to at least 18 inches below the sprinkler heads.

24. (C) HADs (heat actuating devices) are installed along with sprinkler heads on pre-action systems, deluge systems, and combined dry pipe and pre-action systems. These HADs detect the heat rise and fill the system with water before the sprinkler heads open in the pre-action and combined dry pipe and pre-action systems, and open the valve to deliver water to all heads in a deluge system.

25. (D) The primary method used to protect sprinkler heads in areas where the heads are subjected to severe corrosive conditions is to completely coat the head with wax. The wax has a melting temperature slightly below the temperature rating of the heads.

26. (B) A gren-gun is a portable fire extinguisher that fires projectiles filled with dry chemical. It has an advantage over other portable extinguishers of having an extended range and the capability for a fast attack on the fire. The gun fires the projectiles using air pressure. It has the capability of handling four grenades carrying nine pounds of dry powder each.

27. (D) The 165°F head would be uncolored. The 212°F head would have the frame arms painted white, and the 286°F head would have the frame arms painted blue.

28. (B) The question partially describes a pre-action sprinkler system. HADs (heat actuating devices) are used in conjunction with the sprinkler heads and are located in close proximity to the heads. When a

HAD detects a fire, a warning is sounded throughout the building and the system fills with water. The HAD's are set at a temperature rating below that of the heads. The alarm sounding gives occupants a chance to hit the fire before a head ruptures; however, once the heat from the fire reaches the temperature rating of the heads, the heads will go off and place water on the fire. These systems are used in areas where the contents are particularly susceptible to water damage. In the event a head is accidentally knocked off, there will be no water flow.

29.   (A)   The wet pipe system is by far the most common kind used throughout the United States. It represents about 75% of the systems in the nation. In some parts of the country, such as California, it accounts for at least 95% of the systems.

30.   (A)   For a number of years, the primary reason fires have not been controlled or extinguished in buildings having sprinkler systems is that the sprinkler system was shut off at the time of the fire. The other answers given have also been responsible for the system failing to control the fire; however, they have not been as significant as the system being shut off.

Communication Systems

1. There are basically two kinds of municipal fire systems, based upon whether the alarms received at the communication center are

   (A) telephone or telegraph.
   (B) retransmitted manually or automatically.
   (C) radio or telegraph.
   (D) local or radio.

2. What is the largest number of alarm boxes that should be placed on one circuit of a fire alarm system if aerial wire is used in the circuit?

   (A) ten
   (B) fifteen
   (C) twenty
   (D) twenty-five

3. It is recommended that a distinctive colored light be mounted on a fire alarm box or in close proximity to the box in all

   (A) mercantile districts.
   (B) manufacturing districts.
   (C) mercantile and manufacturing districts.
   (D) districts.

4. More than on any other factor, the successful use of the fire alarm system by the public depends on

   (A) the training provided to the public.
   (B) an adequate number of properly spaced and identified boxes.
   (C) the proper identification of box locations.
   (D) the kind of alarm system installed.

5. Type A alarm systems are recommended for communities whenever the number of box alarms received in a year exceeds

   (A) 1,000
   (B) 1,500
   (C) 2,000
   (D) 2,500

6. There are two basic telephone alarm systems currently in service. The difference between the two is based upon

   (A) parallel or series operation.
   (B) grounded or non-grounded operation.
   (C) interfering or non-interfering operation.
   (D) manual or automatic operation.

7. Of the following, the most desirable location for the fire alarm communication center would be in

   (A) a fire station.

(B)   the basement of city hall.
(C)   a separate building in a park.
(D)   a private residence.

8.   The NFPA communication standards

(A)   give more credit for a telegraph-type alarm system.
(B)   give more credit for a telephone-type alarm system.
(C)   give more credit for a radio-type alarm system.
(D)   make no distinction between a telegraph-type, a tele-
        phone-type, and a radio-type alarm system.

9.   If a fire alarm system is properly designed, a person should not
have to travel from the entrance of a dwelling in a congested residential district more than a certain distance to reach an alarm
box.   This distance is

(A)   500 feet.
(B)   one block or 500 feet.
(C)   800 feet.
(D)   two blocks or 800 feet.

10.   The alarm system that sends an alarm directly from private protected property to the municipal fire department is referred to as

(A)   an auxiliary system.
(B)   a remote station system.
(C)   a proprietary system.
(D)   a central station system.

Answers

1. (B)   Standards for fire service communication systems are established by the NFPA.   The two systems are based upon whether the alarms received at the communication center are retransmitted to the fire stations manually or automatically.   With Type A systems, the alarms are retransmitted manually.   With Type B systems, the alarms are retransmitted automatically.

2. (C)   If aerial wire is used in a box circuit, the area protected by the circuit should not be greater than that which could be protected by twenty properly spaced boxes.   If the circuit is entirely underground cable, the area may be extended to that covered by thirty properly spaced boxes.

3. (C)   These lights are recommended to prevent the possibility of delayed alarms.   The lights should be located so that they are clearly visible from all directions for a distance of at least 1,500 feet.

4. (B)   For successful use of a fire alarm system, there should be an adequate number of properly spaced, easily identified boxes.   If boxes are sparse or not properly identified, the chance for successful use of the system is questionable.

5. (D)   A Type A system is recommended whenever the number of box alarms in a community exceeds 2,500 in one year.   Type B systems are designed for communities not desiring a Type A system and when the number of box alarms received in one year is less than 2,500.

6. (A)   The two basic kinds are based upon either a parallel or series operation.   With the parallel operation, each fire alarm box is served by an individual pair of wires.   In the series system, several boxes are served by one circuit.

7. (C)   Fire alarm centers are found in private residences, fire stations, and in city halls; however, the most desirable situation is where they are located in a specially constructed building in a park or other open space.

8. (D)   No distinction is made between the three systems as they all serve the objective of transmitting the alarm from the street to the communication center and from the communication center to the fire stations.

9. (D)   A properly designed system is one where a person in a mercantile or manufacturing district should not have to travel more than one block or 500 feet from the main entrance to the building to reach a box.   In a congested residential district, a box should be available within two blocks or 800 feet from the main entrance of every dwelling.

10. The auxiliary system sends the alarm directly to the municipal fire department.   The remote station system registers the alarm in the office of an agency usually located some distance from the protected property.   A proprietary system receives the alarm at a central supervisory station where trained people are on duty to follow up

with necessary action. A central station system registers the alarm in the office of an independent agency where trained operators are on duty to receive the signal and retransmit it on to the fire department. These central stations are usually located some distance from the protected property.

Fire Extinguishers

1. It is important that the dry chemical used in multipurpose extinguishers not be hygroscopic. This most nearly means that it should not

   (A) be toxic.
   (B) readily take up and maintain moisture.
   (C) be corrosive.
   (D) deplete the oxygen in the air.

2. The NFPA recommends that portable soda-acid and portable foam fire extinguishers be given a standard hydrostatic pressure test every

   (A) year.
   (B) two years.
   (C) five years.
   (D) ten years.

3. There are a number of dry chemicals available for use in fire extinguishers. Of the following kinds, which one is the most corrosive?

   (A) ammonium phosphate base
   (B) potassium chloride base
   (C) sodium bicarbonate base
   (D) monoammonium phosphate base

4. An officer ordered his firefighters to use carbon dioxide extinguishers to attack a fire involving a pile of magnesium chips. This action was

   (A) not correct; this kind of fire cannot be extinguished by smothering.
   (B) correct; this kind of fire can only be extinguished by smothering.
   (C) not correct; the carbon dioxide will intensify the fire.
   (D) correct; this kind of fire can only be extinguished by cooling and since water may not be applied, carbon dioxide is next best.

5. According to NFPA recommendations, a 2½-gallon water extinguisher should not be mounted more than how many feet above the floor?

   (A) 3 feet
   (B) 3½ feet
   (C) 4 feet
   (D) 5 feet

6. A carbon-dioxide extinguisher discharges a stream of gas, rather than a liquid; therefore, it will normally have a range that is

   (A) greater than that of a foam extinguisher.
   (B) greater than that of a soda-acid extinguisher.
   (C) greater than any kind of extinguisher that discharges

       a liquid.
- (D) much less than that of either the foam or soda-acid extinguisher.

7. Markings for extinguishers indicating classes of fires on which they may be used are identified by the letters A, B, C, and D. The markings are also colored. The color blue would indicate that the extinguisher may be used on a

- (A) Class A fire.
- (B) Class B fire.
- (C) Class C fire.
- (D) Class D fire.

8. The chief advantage of a soda-acid extinguisher over a water extinguisher of the same size is that the soda-acid extinguisher

- (A) has greater extinguishing properties.
- (B) is more effective than water on electrical fires.
- (C) supplies a greater volume of extinguishing agents.
- (D) develops its own pressure.

9. Extinguishers are rated by both a letter and a number. The letter indicates the kind of fire the extinguisher may be used on; the number indicates the amount of fire that can be extinguished. Extinguisher ratings for how many of the following kinds of fires may have a number as well as a letter: Class A, Class B, Class C, Class D?

- (A) one
- (B) two
- (C) three
- (D) four

10. Of the following, the most accurate statement concerning the foam fire extinguisher is that

- (A) the operator can control the pressure at which the stress flows from the extinguisher.
- (B) the operator can control the velocity and duration of flow from the extinguisher.
- (C) the extinguisher must be protected from low temperatures or it will freeze.
- (D) gas is present under pressure in the cylinder at all times.

11. Markers are placed on extinguishers to identify on what kind of fire the extinguisher may be used. If placed on the wall near the extinguisher, from what distance should these markers be readable?

- (A) 10 feet
- (B) 15 feet
- (C) 20 feet
- (D) 25 feet

12. Of the following, the <u>least</u> accurate statement concerning the carbon dioxide fire extinguisher is that it

    (A)   is effective on electrical fires.
    (B)   acts to smother the fire.
    (C)   contains gas at high pressure.
    (D)   must be protected from freezing.

13. Which of the following extinguishers was not manufactured after 1969?

    (A)   soda-acid
    (B)   foam
    (C)   cartridge-operated water
    (D)   The manufacture of all the above extinguishers was discontinued in 1969.

14. The 20-pound size carbon dioxide fire extinguisher is recommended for use in

    (A)   electrical power stations.
    (B)   deep-seated fires of ordinary combustible materials such as wood or paper.
    (C)   deep-seated fires of ordinary combustible materials such as textiles or rubbish.
    (D)   none of the foregoing.

15. Which of the following materials would most likely be found in a multipurpose extinguisher?

    (A)   monoammonium
    (B)   sodium bicarbonate
    (C)   potassium bicarbonate
    (D)   bromtrifluoromethane

Answers

1. (B) Hygroscopic refers to the ability to take up and maintain moisture. A hygroscopic dry chemical would clog easily and would not be suitable for use in a fire extinguisher.

2. (C) The hydrostatic pressure test for fire extinguishers recommended by the NFPA is for the purpose of ensuring that the extinguisher is safe to use. The NFPA recommends that portable soda-acid and foam extinguishers be tested every five years. Different testing periods are recommended for other kinds of extinguishers. In addition to the periodic test, a hydrostatic test should be conducted on any extinguisher immediately upon discovering any indication or mechanical injury or corrosion to the extinguisher shell.

3. (B) The residue from potassium chloride extinguishers is somewhat more corrosive than other dry chemicals. On the other hand, the monoammonium phosphate base dry chemical is more difficult to remove from the surfaces where it lands as it hardens when cooled.

4. (C) Extinguishing magnesium fires requires the use of special extinguishing agents. None of the inert gases are effective on magnesium chips. The affinity of magnesium for oxygen is so great that it will burn in an atmosphere of carbon dioxide. Consequently, using a carbon dioxide extinguisher on a fire in a pile of magnesium chips will intensify the fire.

5. (D) The standards recommend that extinguishers having a gross weight not exceeding 40 pounds be installed so that the top of the extinguisher is not more than 5 feet above the floor. Those weighing more than 40 pounds should not be installed with the top more than 3½ feet above the floor. A minimum clearance of four inches should also be maintained between the bottom of the extinguisher and the floor.

6. (D) Of all the portable extinguishers available for use on fires, a closer approach has to be made to the fire when using a carbon dioxide extinguisher than with any other extinguisher. Water and foam extinguishers have a range of 30 to 40 feet while the 5# carbon dioxide extinguisher has a range of about 3 feet and the 20-pound size has a range of about 7 feet.

7. (C) The markings are as follows:

    Class A--a green triangle
    Class B--a red square
    Class C--a blue circle
    Class D--a yellow five-pointed star

8. (D) It is assumed from the question that the other water extinguisher is a hand-pump type. In this case, the soda-acid extinguisher would have an advantage over a pump type extinguisher in the hands of an inexperienced person as it would develop its own pressure. There is, however, a chance that the soda-acid extinguisher would fail to work more often than a hand-pump water extinguisher. (Note: Soda-acid extinguishers have not been manufactured since 1969. How-

ever, they are still found in various locations making it necessary for firefighters to be familiar with their operation.)

9. (B) Ratings for extinguishers for Class A and Class B fires also have a number, as an example, A-2 or A-2, B-4. Ratings for Class C and Class D extinguishers do not have a number.

10. (C) Portable foam extinguishers contain 2½ gallons of water which is mixed with bicarbonate of soda. The water must be protected against freezing in those areas subject to low temperatures. (Note: Foam extinguishers have not been manufactured since 1969. However, they will still be found in various locations making it necessary for firefighters to be familiar with their operation.)

11. (D) These markings may be found on the extinguisher, on the wall near the extinguisher, or in both locations. If placed on the extinguisher, they should be readable from a distance of 3 feet. If placed on the wall near the extinguisher, they should be readable from a distance of 25 feet.

12. (D) Contrary to what many people think, the cooling effect of carbon dioxide from an extinguisher is practically nil. All the answers given, except (D), however, are true regarding carbon dioxide extinguishers.

13. (D) The best way to remember this is that all inverter-type (those requiring that the extinguisher be turned upside down to operate) extinguishers were discontinued in 1969. These extinguishers, however, will still be found in many locations, making it necessary that firefighters understand their principle and method of operation.

14. (A) A fire in energized electrical equipment is classified as a Class C fire. Carbon dioxide extinguishers are rated for use on Class C fires due to the nonconductivity of the carbon dioxide.

15. (A) Monoammonium is a dry chemical that is used in multipurpose extinguishers. One of the primary reasons for its use is that it is considerably less hygroscopic than some of the original powders used in multipurpose extinguishers.

VII

EMERGENCY MEDICAL CARE

Questions 1 through 15 in this chapter are based upon information contained in the text <u>First Responder</u> (J. David Bergeron, Robert J. Brady Company, Bowie, Md., 1982). This information has been used with the expressed permission of the Robert J. Brady Company.

1. You respond to an incident and find a man who is conscious and clear of mind. He is bleeding badly from his right arm. The man tells you that he does not want you to treat him. In this instance

   (A) you must force treatment on him as failure to assist
        may be considered abandonment.
   (B) the man has the right to refuse your assistance.
   (C) you have the choice of either forcing treatment or
        not treating.
   (D) you may only treat the man if a law enforcement offi-
        cial is on the scene and instructs you to proceed.

2. When arriving at the scene of a medical emergency and examining the patient, your first concern is to

   (A) stabilize the patient.
   (B) identify and correct life-threatening problems.
   (C) keep the patient stable and continue to monitor the
        patient in case the condition worsens.
   (D) identify any injuries or medical problems and correct
        these problems.

3. The purpose of the primary survey of a patient at the scene of a medical emergency is to

   (A) detect life-threatening problems.
   (B) determine if the patient is suffering from any
        disease.
   (C) identify any injuries.
   (D) stabilize the patient.

4. A person is removed from a smoke-filled area of a burning building. When examining him, you notice that his skin is cherry red. This is an indication that the possible problem is probably

   (A) a heart attack.

(B) high blood pressure.
(C) carbon monoxide poisoning.
(D) insulin shock.

5. All of the following can be considered vital signs for a First Responder to a medical emergency except

(A) pulse.
(B) respiration.
(C) skin temperature.
(D) condition of the eyes.

6. It can be considered that most adults at rest have a respiration rate of between

(A) 12 and 15 breaths per minute.
(B) 10 and 14 breaths per minute.
(C) 15 and 20 breaths per minute.
(D) 16 and 22 breaths per minute.

7. A pedal pulse of a patient is taken on the

(A) wrist.
(B) arm.
(C) throat.
(D) foot.

8. When examining a patient, cool, clammy skin would most likely indicate

(A) loss of body heat.
(B) exposure to cold.
(C) shock.
(D) infection.

9. When examining a patient, you notice that the pupils of his eyes are unequal in size. This would most likely indicate

(A) shock.
(B) certain drug abuses.
(C) stroke or head injury.
(D) a coma.

10. When examining a person for a possible partial obstructed airway you note that the patient is breathing with a gurgling sound. This most nearly means that the problem is

(A) caused by the tongue obstructing the back of the throat.
(B) caused by a foreign object in the windpipe or by blood in the airway.
(C) caused by spasms of the voicebox.
(D) due to swellings or spasms along the lower airway.

11. If a person with an obstructed airway is standing or sitting, it is best to deliver how many sharp blows to the midline of the patient's

back, in the area between the shoulder blades?

(A) one
(B) two
(C) three
(D) four

12. How many breaths per minute should be delivered to an adult person when performing mouth-to-mouth ventilation?

(A) ten
(B) twelve
(C) fifteen
(D) twenty

13. The CPR compression site for an adult is located

(A) directly over the center of the heart.
(B) directly over the xiphoid process.
(C) two finger widths above the xiphoid process.
(D) two finger widths below the xiphoid process.

14. When applying CPR to an adult male, compressions must be delivered at a rate of how many per minute?

(A) 50
(B) 60
(C) 70
(D) 80

15. When someone stops breathing and the heart stops beating, clinical death results. Biological death occurs when the brain cells start to die, within how many minutes?

(A) 3 to 5 minutes
(B) 4 to 6 minutes
(C) 5 to 7 minutes
(D) 6 to 8 minutes

16. The best first-aid treatment for an extensive and somewhat severe burn caused by chemicals is

(A) applying soda compresses.
(B) thoroughly washing with water.
(C) applying tannic acid.
(D) applying a weak vinegar solution.

17. Which of the following statements regarding the first aid treatment of a second-degree burn is true?

(A) Immerse the burned part in ice water until the burn subsides.
(B) Break the blisters.
(C) Use an antiseptic preparation on a severe burn.
(D) If the arms or legs are affected, keep them elevated.

18. First-aid treatment for frostbite should include

    (A) gentle massage of the affected part.
    (B) rubbing the affected part with snow.
    (C) applying a hot-water bottle close to the affec-
        ted part.
    (D) bringing the patient into a warm room as soon as
        possible.

19. A worker is unconscious from electric shock and presents a blue
    appearance. You should

    (A) keep the victim quiet and apply artificial respiration.
    (B) apply cold applications and try to rouse the victim.
    (C) lay the victim down with head slightly raised and give
        cold applications.
    (D) apply external warmth and try to rouse the victim.

20. Of the following methods of controlling severe bleeding, the pre-
    ferred method is

    (A) direct pressure.
    (B) elevation.
    (C) pressure on the supplying artery.
    (D) tourniquet.

21. If it becomes necessary to apply a tourniquet as a first-aid measure,
    which of the following procedures is proper?

    (A) Use anything available, such as rope, sash cord, or
        wire.
    (B) Attach a notation to the patient giving location and
        hour of application.
    (C) Place a tourniquet at the edge of the wound and loosen
        it every 15 to 20 minutes.
    (D) First place a bandage over the wound and then apply the
        tourniquet over the bandage.

22. Which of the following is one of the correct first-aid steps when
    treating a person for heat exhaustion?

    (A) Take the victim into an air-conditioned room, if
        possible.
    (B) Give the victim sips of salt water over a period of
        about one hour.
    (C) Have the victim lie down and raise his head from 8
        to 12 inches.
    (D) Loosen the victim's clothing and apply warm, dry
        cloths to his body.

23. Apply cold, wet packs or place a small bag of crushed ice on the
    affected area, over a thin towel to protect the victim's skin. The
    above statement applies to the first-aid treatment for

    (A) a dislocation.
    (B) a sprain.

(C)   a strain
(D)   all of the above

24.   The general principle in cases of shock is to use external heat on a patient

(A)   at all times.
(B)   freely.
(C)   usually.
(D)   with caution.

25.   At the scene of an automobile accident, you find a victim with an open wound of the abdomen with the intestines protruding. Which, if any, of the following are incorrect procedures to use for the first-aid treatment of this victim?

(A)   Give the victim a cool drink of water.
(B)   Push the protruding intestines back into the abdom-
         inal cavity.
(C)   If breathing is difficult, keep the victim's feet
         elevated.
(D)   All of the above are incorrect procedures.

26.   While on-duty at a fire in a remote area of the city, one of your firefighters gets a large amount of caustic acid in one eye. You should first

(A)   bandage the eye.
(B)   notify the signal office.
(C)   put a few drops of mineral oil in the eye.
(D)   irrigate the eye with a large quantity of water.

27.   Blisters can be caused by friction from shoes or boots and may appear on the heels, toes, or tops of the feet. In the first-aid treatment for these blisters, they

(A)   should not be broken open.
(B)   should be broken open.
(C)   may or may not be broken, depending on the situation.

28.   Which of the following is the worst danger to which puncture wounds are subject?

(A)   gangrene
(B)   tetanus
(C)   hemorrhage
(D)   acne

29.   The proper first-aid treatment for a tick bite is to

(A)   put alcohol on the bite and cover it with a sterile
         bandage.
(B)   place a hot cigarette next to the tick but do not
         touch the skin with the cigarette.
(C)   cover the tick with heavy oil to close its breathing
         pores.

(D)   soak the affected part in cold water.

30.   When the unconscious victim does not breathe, or when the victim appears to be breathing, but the air passageway is blocked, the brain dies from oxygen lack within a

(A)   matter of hours.
(B)   very few minutes.
(C)   few days
(D)   very few seconds.

Answers

1. (B) Adults, when conscious and clear of mind, have the right to refuse your care. Their reasons may be religious or they may base their decision on a lack of trust. They may have reasons that you find senseless. For whatever reasons, a competent adult may refuse care. You cannot force care upon them, nor can you legally restrain them until EMTs arrive. Your only course of action is to try to gain their confidence through conversation. This will help if fear is the reason why your help was refused. Do not argue with a patient, particularly if his reasons are based on religious beliefs. The added stress from the argument could cause serious complications.

2. (B) Your primary concern is to identify and correct life-threatening problems. Your second concern is to identify any injuries or medical problems and correct these problems or stabilize the patient. Your third concern is to keep the patient stable and to continue to monitor the patient in case the condition worsens.

3. (A) The primary survey is defined as a process carried out in order to detect life-threatening problems. The three major life-threatening problems are: 1) respiration; 2) circulation; 3) bleeding.

4. (C) The following chart is a good reference to use to determine a possible cause of a problem:

Observation	Possible Problem
red skin	high blood pressure, heart attack, heat stroke, diabetic coma
cherry red skin	carbon monoxide poisoning
pale, ashen skin	shock, heart attack, bleeding, fright, insulin shock
blue skin	heart failure, airway problems, lung disease, certain poisonings

5. (D) The vital signs are: pulse, respiration, and skin temperature. Vital signs tell much about the stability of the patient. These signs can also alert a first-aider to problems that require immediate attention.

6. (A) Normal respiration rates for adults at rest fall into a range of 12 to 20 breaths per minute, with most people having 12 to 15 breaths per minute as their normal range. Older adults tend to breathe more slowly than young adults. A rate over 28 breaths per minute is serious.

7. (D) The circulation of blood through the leg to the foot can be confirmed by feeling a pedal (PED-al) pulse. This pulse can be felt between the big toe and the second toe.

8. (C) The following chart can be used as a guide to problems indicated by skin temperature:

Observation	Possible Problem
cool, moist skin	shock, bleeding, loss of body heat, heat exhaustion
cool, dry skin	exposure to cold
cool, clammy skin	shock
hot, dry skin	high fever, heat stroke
hot, moist skin	infection

9. (C) The following chart can be used as a guide to problems indicated by the pupils of the eyes:

Observation	Possible Problem
dilated, unresponsive	unconsciousness, shock, cardiac arrest, bleeding, certain drug abuses
constricted, unresponsive	central nervous system damage, certain drug abuses
unequal pupils	stroke, head injury
lackluster	shock, coma

10. (B) Following are the indications of atypical breathing sounds:

Snoring--probably caused by the tongue obstructing the back of the throat.
Gurgling--probably caused by a foreign object in the windpipe, or by blood in the airway.
Crowing--probably caused by spasms of the voicebox.
Wheezing--may be due to swellings or spasms along the lower airway. However, wheezing often does not indicate any major problem of the airway.

11. (D) If a person with an obstructed airway is standing or sitting, the proper procedure to use to help clear the airway is:

1. Stand slightly behind the patient, at the side.
2. Place one hand on the middle of the patient's chest at the level of the collarbones.
3. Use this hand for support as you bend the patient at the waist.
4. Bend the patient forward until the head is at or below the chest level.
5. With the heel of your free hand, deliver four sharp blows to the midline of the patient's back, in the area between the shoulder blades. These blows must

be delivered forcefully, in rapid succession.

12. (B) For the mouth-to-mouth ventilation technique, you must deliver breaths to the patient at <u>one breath every five seconds</u>, or at a rate of <u>twelve breaths per minute</u>. You will have to continue at this rate until the patient begins to breathe unaided, until someone trained in mouth-to-mouth techniques can replace you, or until you are too exhausted to continue.

13. (C) In order to locate the CPR compression site, begin by finding the lower border of the breastbone. You will find an area where the hard breastbone stops and the soft abdomen begins. In this region there is a small structure extending downward from the breastbone. This is called the xiphoid process. Once the xiphoid process has been located, the CPR compression site may be found by measuring two finger widths toward the patient's head, staying along the centerline of the breastbone.

14. (D) The rate at which both compressions and ventilations are delivered is critical in providing CPR. To provide proper assistance, you must know the following:

1. Compressions must be delivered at a rate of 80 per second.
2. Ventilations must be delivered at a rate of two breaths every 15 compressions. These two breaths must be delivered as quickly as possible.
3. CPR should never be interrupted for more than 5 seconds.

When compressions are delivered at a rate of 80 per minute, 80 compressions are not delivered per minute. The time taken up for ventilations will mean that about 60 compressions per minute are actually being delivered to the patient.

15. (B) There is a strong relationship between breathing, circulation, and brain activity. If one of the three stops, all three will stop. When a patient's heart stops beating, the patient is in cardiac arrest. The patient will be unconscious and have a "death-like" appearance on the face.

16. (B) The first-aid treatment for chemical burns of the skin is to wash away the chemical with large quantities of water as quickly as possible. A shower or garden hose should be used, if available. Washing should continue for at least five minutes. It is important that the victim's clothing be removed from the area or areas involved.

17. (D) The burned part should be immersed in cold water, not ice water. The blisters should not be broken, and an antiseptic preparation should not be used on a severe burn. However, the arms or legs should be kept elevated if they were affected.

18. (D) The objectives of first-aid treatment for a frostbite are to protect the frozen area from further injury, to warm the affected area rapidly, and to maintain respiration. The person should be taken into a warm room as soon as possible.

19. (A) The condition of the victim given in the question indicates that artificial respiration should be started immediately. Recovery will probably be slow, and it will be necessary to continue the treatment for a long time. The treatment should be continued until one of the following occurs:

1. The victim begins to breathe.
2. The victim is pronounced dead by a doctor.
3. The victim is dead beyond any doubt.

20. (A) Direct pressure by a hand over a dressing is the preferred method since it prevents the loss of blood without interfering with the normal blood circulation.

21. (B) None of the answers given except (B) should be performed when a tourniquet is applied. Tourniquets should only be used as a last resource and, when used, amount to a decision to risk the sacrifice of a limb in order to save a life. A note should be attached to the victim's clothing giving the location of the tourniquet and the time it was applied.

22. (C) All the answers except (C) are incorrect. The salt-water ratio should be one teaspoonful of salt per glass, one-half a glass every 15 minutes.

23. (B) Neither heat nor cold is recommended to be used on a dislocation. Strains require the application of heat; warm, wet applications. The treatment given in the question is for a sprain.

24. (B) Shock is the failure of the cardiovascular system to provide sufficient blood circulation to every part of the body. One treatment for shock is to prevent the loss of body heat. A blanket should be placed over the patient and also under the patient if the patient is on the ground. Although it is important to prevent the loss of body heat, it is also important not to overheat the body.

25. (D) The victim should not be given any fluids or solid food because surgery will be necessary. Do not try to push the protruding intestines back into the abdominal. If breathing is difficult, keep the victim's head and shoulders elevated with a pillow or folded coat.

26. (D) The eye should be flooded with water for at least 15 minutes. If the person is lying down, the head should be turned to the side and the lid held open. The water should be poured from the inner corner outward. Any loose particles of the acid should be lifted off with a sterile gauze or a clean handkerchief. After thoroughly flushing with water, immobilize the eye by covering it with a dry pad or protective cloth and seek immediate medical attention.

27. (C) If all pressure can be relieved until the fluid is absorbed, blisters are best left unbroken. Otherwise, wash the entire area with soap and water and make a small puncture hole at the base of the blister with a needle that has been sterilized in a match flame or by soaking in rubbing alcohol.

28.   (B)   A puncture wound is caused by an object piercing the skin layers, causing a small hole in the tissues. External bleeding is usually limited, which increases the hazard of infection. Tetanus may develop from puncture wounds.

29.   (C)   Cover the tick with heavy oil to close its breathing pores. The oil to use is mineral, salad, or machine. This procedure may disengage the tick immediately. If not, allow the oil to remain in place for half an hour, then carefully remove the tick with tweezers, taking care to ensure that all parts are removed.

30.   (B)   Of all the body tissues, the brain is the most sensitive to a lack of oxygen. If the brain does not receive oxygen for a period of 4 to 6 minutes due to stoppage of breathing and no heart beat, the brain will probably be damaged to the extent that a victim may never become conscious, even if breathing is resumed.

STUDY SOURCES

One of the problems confronting a firefighter preparing for promotional examinations is what to study. There are three books that are almost a must: 1) The NFPA Fire Protection Handbook; 2) Municipal Fire Administration; 3) Fire Chief's Handbook.

There are a multitude of books available other than these three; however, it is almost impossible to study all of them. Consequently, those preparing for promotion must select a few, or take the information deemed necessary from a number of books. The Standards published by the National Fire Protection Association (NFPA) are useful as is the series of manuals published by the International Fire Service Training Association (IFSTA). These publications will be referred to as the NFPA Standard and the IFSTA numbers in this chapter. In addition, the following constitute recommendations covering the subject matter included in this manual.

Supervision

Bryan, John L., and Raymond C. Picard, eds., Managing Fire Services, International City Managers' Association, 1979.

Coleman, Ronny J., Management of Fire Service Operations, Duxbury Press, 1978.

Effective Supervisory Practices. International City Managers' Association, 1978.

George, Claude S., Supervision in Action: The Art of Managing Others, Reston Publishing Co., 1979.

Gratz, David B., Fire Department Management: Scope and Method, Glencoe Press, 1972.

Sayles, Leonard, and George Strauss, Managing Human Resources, Prentice-Hall, 1981.

Spriegel, William R., and Edward Schulz, Elements of Supervision, John Wiley & Sons, 1957.

Training

Fire Service Instructor, IFSTA 303.

Granito, Anthony R., Fire Instructor's Training Guide, Dun-Donnelley Publishing Corp., 1972.

Public and Community Relations

Cameron, Cyril T., <u>Public Relations in the Emergency Department</u>, Robert J. Brady Co., 1980.

Gilbert, William H., <u>Public Relations in Local Government</u>, International City Managers' Association, 1975.

Reilly, R., <u>Public Relations in Action</u>, Prentice-Hall, 1981.

<u>Public Fire Education</u>, IFSTA 606.

Safety

<u>Fire Department Safety Officer</u>, NFPA #1501.

<u>Firefighter Occupational Safety</u>, IFSTA 209.

<u>Safety in the Fire Service</u>, NFPA.

Firefighting—Principles and Procedures

Clark, William E., <u>Fire Fighting Principles and Practices</u>, Fire Engineering, 1974.

Fried, Emanuel, <u>Fireground Tactics</u>, H. Marvin Ginn Corporation, 1972.

<u>Fire Attack 1</u>, Kimbal, Warren Y. NFPA, 1966.

--<u>Fire Attack 2</u>, NFPA, 1968.

Layman, Lloyd L., <u>Attacking and Extinguishing Interior Fires</u>, NFPA, 1960.

Walsh, Charles V., and Leonard G. Marks, <u>Firefighting Strategy and Leadership</u>, McGraw-Hill Book Co., 1977.

Shipboard Fires

<u>Marine Fire Prevention, Firefighting and Fire Safety</u>, Marine Training Advisory Board, Robert J. Brady Co.

Hazardous Materials Fires and Emergencies

See the section on <u>Hazardous Materials</u>.

Wildfires

Gaylor, Harry P., <u>Wildfires--Prevention and Control</u>, Robert J. Brady Co., 1974.

<u>Ground Cover Fire Fighting Practices</u>, IFSTA 207.

Satterlund, Donald R., <u>Wildland Watershed Management</u>, John Wiley and Sons, 1972.

Flammable Liquid and Gas Fires

See the section on <u>Hazardous Materials</u>.

High-Rise Building Fires

<u>Fire Problems in High-Rise Buildings</u>, IFSTA 304.

Fried, Emanuel, <u>Fireground Tactics</u>, H. Marvin Ginn Corp., 1972.

<u>High-Rise Building Fires and Fire Safety</u>, NFPA, 1972.

Mendes, Robert F., <u>Fighting High Rise Building Fires--Tactics and Logistics</u>, NFPA, 1975.

O'Hagan, John T., <u>High Rise Fire and Life Safety</u>, 1977.

Schueller, Wolfgang, <u>High Rise Building Structures</u>, Wiley-Interscience, 1977.

Aircraft Fires

<u>Aircraft Fire Protection and Rescue Procedures</u>, IFSTA 206.
<u>Standard Operating Procedures, Aircraft Rescue & Fire Fighting</u>, NFPA
   #402.

Pumps and Pumping Equipment

<u>Automotive Fire Apparatus</u>, NFPA #1901.
Casey, James F., <u>Fire Service Hydraulics</u>, Dun-Donnelley Publishing Corp.,
   1970.
Erven, Lawrence W., <u>Fire Fighting Apparatus and Procedures</u>, Glencoe
   Publishing Co., 1979.
Isman, Warren, <u>Fire Service Pumps and Hydraulics</u>, Delmar Publishing Co.,
   1977.
Mahoney, Gene, <u>Introduction to Fire Apparatus and Equipment</u>, Allyn and
   Bacon, Inc., 1981.

Hydraulics

Casey, James F., ed., <u>Fire Service Hydraulics</u>, Dun-Donnelly Publishing
   Corp., 1970.
Erven, Lawrence W., <u>Techniques of Fire Hydraulics</u>, Glencoe Press, 1972.
Isman, Warren, <u>Fire Service Pumps and Hydraulics</u>, Delmar Publishing Co.,
   1977.
Mahoney, Eugene F., <u>Fire Department Hydraulics</u>, Allyn and Bacon, Inc.,
   1980.
Purington, Robert C., <u>Firefighting Hydraulics</u>, McGraw-Hill, 1974.

Fire Apparatus

<u>Automotive Fire Apparatus</u>, NFPA #1901.
Erven, Lawrence W., <u>Fire Fighting Apparatus and Procedures</u>, Glencoe Pub-
   lishing Co., 1979.
Mahoney, Gene, <u>Introduction to Fire Apparatus and Equipment</u>, Allyn and
   Bacon, Inc., 1981.

Water Supply

Cozad, F. Dale, <u>Water Supply for Fire Protection</u>, Prentice-Hall, 1981.
Mahoney, Eugene F., <u>Fire Department Hydraulics</u>, Allyn and Bacon, Inc.,
   1980.
<u>Water Supplies for Fire Protection</u>, IFSTA 205.

Fire Prevention—Practices and Procedures

Bare, William K., <u>Fundamentals of Fire Prevention</u>, John Wiley & Sons,
   1977.
Clet, Vince H., <u>Fire-Related Codes, Laws and Ordinances</u>, Glencoe Publishing Co., 1978.
<u>Fire Prevention and Inspection</u>, IFSTA 110.
Robertson, James C., <u>Introduction to Fire Prevention</u>, Glencoe Publishing Co., 1979.
Whitman, Lawrence, <u>Fire Prevention</u>, Nelson-Hall, 1979.

Hazardous Materials

1980 Emergency Response Guidebook--Hazardous Materials, Department of Transportation.
Fire Protection Guide on Hazardous Materials, NFPA.
Isman, Warren E., and Gene P. Carlson, Hazardous Materials, Glencoe Publishing Co., 1980.
Meidl, James H., Explosives and Toxic Materials, Glencoe Press, 1980.
Meidl, James H., Flammable Hazardous Materials, Glencoe Press, 1970.
Schieler, Leroy, and Denis Pauze, Hazardous Materials, Delmar Publishers, 1976.

Arson

Bahme, Charles W., Fire Service and the Law, NFPA, 1976.
Battle, Bredan P., Arson Detection and Investigation, Arco Publishing Co., 1978.
Carter, Robert E., Arson Investigation, Macmillan Publishing Co., 1978.
French, Harvey, The Anatomy of Arson, Arco Publishing Co., 1979.
Kirk, Paul L., Fire Investigation, John Wiley & Sons, 1969.
Roblee, Charles L., and Allen J. McKechnie, The Investigation of Fires, Prentice-Hall, 1981.

Building Construction

Brannigan, Francis L., Building Construction for the Fire Service, NFPA, 1971.

Fire Protection Systems

Averill, C.F., Sprinkler Systems Designs: Past, Present & Future, Society of Fire Protection Engineers, 1979.
Bryan, John L., Fire Suppression and Detection Systems, Macmillan Publishing Co., 1974.

Communication Systems

Auxiliary Protective Signaling Systems, NFPA #72B.
Central Station Signaling Systems, NFPA #71.
Local Protective Signaling Systems, NFPA #72A.
Proprietary Protective Signaling Systems, NFPA #72D.
Public Fire Service Communications, NFPA #1221.
Remote Station Protective Signaling Systems, NFPA #72C.

Fire Extinguishers

Portable Fire Extinguishers, NFPA #10.

Emergency Medical Care

Barber, Janet M., and Peter A Dillman, Emergency Patient Care for the EMT-A, Reston Publishing Co., 1981.
Bergeron, J. David, First Responder, Robert J. Brady Co., 1982.
Erven, Lawrence W., Emergency Rescue, Glencoe Publishing Co., 1980.
Gazzaniga, Alan B., Lloyd T. Iseri, and Martin Baren, Emergency Care:

Principles and Practices for the EMT--Paramedic, Reston Publishing
    Co., 1979.
Grant, Harvey, and Robert Murray, Emergency Care, Robert J. Brady Co.,
    1978.
Huszar, Robert J., Emergency Cardiac Care, Robert J. Brady Co., 1982.
Parcel, Guy S., Basic Emergency Care of the Sick and Injured, C.V. Mosby
    Co., 1982.

MORE ARCO BOOKS

Perhaps you've discovered that you are weak in language, verbal ability or mathematics. Why flounder and fail when help is so easily available? Brush up in the privacy of your own home with one of our review books.

At the same time, choose from our wide range of hobby and general interest books, designed to entertain and inform you in whatever area you select.

Each of the following books was created under the same expert editorial supervision that produced the excellent book you are now using. Whatever your goals or interests. . . you can learn more and score higher on tests with Arco.

HIGH SCHOOL AND COLLEGE PREPARATION

GED PREPARATION

General Education Development Series

COLLEGE BOARD ACHIEVEMENT TESTS/CBAT

ARCO COLLEGE OUTLINES

CIVIL SERVICE AND TEST PREPARATION—GENERAL

Road Car Inspector (T.A.) 03743-1 8.00
Sanitation Foreman (Foreman & Asst. Foreman) 01958-1 6.00
Sanitation Man 00025-2 4.00
School Crossing Guard 00611-0 4.00
Senior Clerical Series 01173-4 8.00
Senior Clerk—Stenographer 01797-X 9.00
Senior File Clerk 00124-0 8.00
Senior and Supervising Parking Enforcement Agent 03737-7 6.00
Senior Typist 03870-5 6.00
Sergeant, P.D. 00026-0 10.00
Shop Clerk 03684-2 6.00
Social Supervisor 04190-0 8.00
Staff Attendant . . . . .LR 01739-2 6.50
Staff Positions: Senior Administrative Associate
    and Assistant 03490-4 6.00
State Trooper 05234-1 8.00
Stenographer—Typist (Practical Preparation) 00147-X 6.00
Stenographer—U.S. Government Positions GS 2-7 . . . . . 04388-1 6.00
Storekeeper—Stockman (Senior Storekeeper) 01691-4 8.00
Structural Apprentice 00177-1 5.00
Structure Maintainer Trainee, Groups A to E 03683-4 6.00
Supervising Clerk (Income Maintenance) 02879-3 5.00
Supervising Clerk—Stenographer 04309-1 8.00
Supervision Course 01590-X 8.00
Surface Line Dispatcher 00140-2 6.00
Tabulating Machine Operator (IBM) 00781-8 4.00
Taking Tests and Scoring High, Honig 01347-8 4.00
Telephone Maintainer: New York City Transit Authority . 03742-3 5.00
Test Your Vocational Aptitude, Asta & Bernbach 03606-0 6.00
Towerman (Municipal Subway System) 00157-7 5.00

Trackman (Municipal Subways) 00075-9 5.00
Track Foreman: New York City Transit Authority 03739-3 6.00
Traffic Control Agent 03421-1 5.00
Train Dispatcher 00158-3 5.00
Transit Patrolman 00092-9 5.00
Transit Sergeant—Lieutenant 00161-5 4.00
Treasury Enforcement Agent 00131-3 8.00
U.S. Postal Service Motor Vehicle Operator 04426-8 8.00
U.S. Professional Mid-Level Positions
    Grades GS-9 Through GS-12 . . . . 02036-9 6.00
U.S. Summer Jobs 02480-1 4.00
Ventilation and Drainage Maintainer: New York City
    Transit Authority 03741-5 6.00
Vocabulary Builder and Guide to Verbal Tests 00535-1 5.95
Vocabulary, Spelling and Grammar 00077-5 5.00
Welder 01374-5 8.00
X-Ray Technician (See Radiologic Technology
    Exam Review) 03833-0 8.00

MILITARY EXAMINATION SERIES

Practice for Air Force Placement Tests 04270-2 6.00
Practice for Army Classification and Placement
    (ASVAB) 03845-4 8.00
Practice for the Armed Forces Tests 05303-8 6.00
Practice for Navy Placement Tests 04560-4 6.00
Practice for Officer Candidate Tests 01304-4 6.00
Tests for Women in the Armed Forces 03821-7 6.00
U.S. Service Academies 01544-6 6.00

ARCO SCHOLARSHIP EXAMINATION SERIES

AP

Advanced Placement Music, Seligson-Ross 04743-7 4.95

AP/CBAT

Advanced Placement and College Board
    Achievement Tests in Physics (B-C),
    Arco Editorial Board 04493-4 6.95

AP/CLEP

Advanced Placement and College Level Examinations in
    American History, Woloch 03804-7 5.95
Advanced Placement and College Level Examinations in
    Biology, Arco Editorial Board 04415-2 5.95
Advanced Placement and College Level Examinations in
    Chemistry 04484-5 4.95
Advanced Placement and College Level Examinations in
    English—Analysis and Interpretation of Literature . . 04406-3 4.95

AP/CLEP/CBAT

Advanced Placement, College Level Examinations and
    College Board Achievement Tests In European
    History 04407-1 5.95

CLEP

College Level Examination in Composition and Freshman
    English 03798-9 4.95

College Level Examination Program 04150-1 6.00

College Level Examination Program:
    The General Examination in the Humanities 04727-5 6.95

College Level Examinations in Mathematics: College
    Algebra, College Algebra-Trigonometry,
    Trigonometry 04339-3 5.95

MEDICINE

MEDICAL REVIEW BOOKS

Basic Dental Sciences Review, DeMarco 03396-7 10.00
Basic Science Nursing Review, Cheatham,
    Fitzsimmons, Lessner, King, Lafferty,
    DePace & Blumenstein 05133-7 8.00
Biochemistry Review, Silverman 04359-8 12.00
Clinical Dental Sciences Review, DeMarco 03383-5 10.00
Comprehensive Medical Boards Examination Review,
    second revised edition, Horemis 01595-0 8.00
Dental Assistants Examination Review, Hirsch 03902-7 9.00
Dental Hygiene Examination Review, Armstrong 04283-4 10.00

Endocrinology Review, Hsu 04228-1 12.00
General Pathology Review, Lewis & Kerwin 04774-7 10.00
Histology and Embryology Review, Amenta 03831-4 8.00
Human Anatomy Review, Montgomery & Singleton 03368-1 8.00
Human Physiology Examination Review, Shepard 04826-3 12.00
Internal Medicine Review, Pieroni 03881-0 11.00
Medical Assistants Examination Review, second edition,
    Clement 04854-9 10.00
Medical Examinations: A Preparation Guide, Bhardwaj . . 03944-2 9.00
Medical Technology Examination Review, Hossaini 04365-2 10.00
Microbiology and Immunology Review, Second edition,
    Rothfield, Ward & Tilton 04882-4 10.00

Neuroscience and Clinical Neurology Review, Goldblatt . 03370-3 10.00
Nuclear Medicine Technology
    Examination Review, Spies 04724-0 12.00
Obstetrics and Gynecology Review, Second edition,
    Vontver 03450-5 9.00
Patient Management Problems:
    Obstetrics and Gynecology, DeCherney 04364-4 8.00
Patient Management Problems:
    Pediatrics, Howell & Simon 04780-1 10.00
Patient Management Problems: Surgery, Rosenberg 04654-6 8.00
Pediatrics Review, Second edition, Lorin 03375-4 8.00
Pharmacology Review, Ellis 04108-0 10.00
Pharmacy Review, Second edition, Singer 04878-6 12.00
Physical Medicine and Rehabilitation
    Review, Schuchmann 04723-2 15.00
Physician's Assistant Examination Review,
    Aschenbrener 04026-2 12.00
Psychiatry Examination Review—Second Edition,
    Easson 03395-9 8.00
Psychiatry: Patient Management Review, Easson 04058-0 8.00
Public Health and Preventive Medicine Review 04690-2 9.00
Pulmonary Disease Review, Hall 04008-4 12.00
Radiologic Technology Examination Review,
    Naidech & Damon 03833-0 8.00
Specialty Board Review: Anatomic Pathology,
    Gravanis & Johnson 03858-6 14.00
Specialty Board Review: Anesthesiology, Beach 04112-9 14.00
Specialty Board Review: Family Practice,
    Bhardwaj & Yen 03943-4 12.00
Specialty Board Review: General Surgery,
    Rob & Hinshaw 03494-7 12.00
Specialty Board Review: Internal Medicine, Pieroni ... 04818-2 14.00
Specialty Board Review: Obstetrics and
    Gynecology, Williams 03477-7 14.00
Specialty Board Review: Psychiatry, Atkins 03471-8 12.00
Surgery Review, Kountz et al 03880-2 8.00
Systemic Pathology Review, Lewis & Kerwin 04930-8 12.00

MEDIBOOKS

Fundamentals of Radiation Therapy, Lowry 03462-9 7.50
Midwifery, Hallum 03460-2 5.50
Pathology, Mayers 04774-7 6.00

Principles of Intensive Care, Emery, Yates & Moorhead . 03461-0 6.00

MEDICAL TEXTBOOKS AND MANUALS

The Basis of Clinical Diagnosis, Parkins & Pegrum 03660-5 12.95
Differential Diagnosis in Gynecology,
    Vontver & Gamette 04129-3 12.00
Differential Diagnosis in Neurology, Smith 04033-5 14.00
Differential Diagnosis in Obstetrics, Williams
    & Joseph 04161-7 10.00
Differential Diagnosis in Disorders of the Eye,
    Kupfer & Kaiser-Kupfer 04315-6 10.00
Differential Diagnosis in Otolaryngology, Lee 04017-2 14.00
Differntial Diagnosis of Renal and Electrolyte Disorders,
    Klahr 04063-7 14.00
The Effective Scutboy, Harrell & Firestein 05159-0 7.50
Emergency Medicine, Hocutt 04983-9 12.00
Hospital-Based Education, Linton & Truelove 04776-3 10.00
Modern Medicine, Read et al 04124-2 14.75
Psychiatry: A Concise Textbook for Primary
    Care Practitioners, Kraft et al 03924-8 12.00
Simplified Mathematics for Nurses,
    McElroy, Carr & Carr 04197-8 5.00

NURSING REVIEW BOOKS

Arco's Comprehensive State Board
    Examination Review for Nurses, Carter 04925-1 8.95
Child Health Nursing Review, Second edition, Porter ... 04825-5 7.00
Fundamentals of Nursing Review, Carter 04512-4 6.00
Maternal Health Nursing Review, Second edition,
    Sagebeer 04822-0 6.00
Medical-Surgical Nursing Examination Review,
    Second edition, Horemis & Matamors 02511-5 6.00
Medical-Surgical Nursing Review, Second edition,
    Hazzard 04823-9 6.00
Nursing Comprehensive Examination Review,
    second revised edition, Horemis 02499-2 6.00
Nursing Exam Review in Basic Sciences,
    Horemis & Matamors 02946-3 4.00
Practical Nursing Review, Second edition, Redempta ... 04827-1 7.50
Practice Tests for the L.P.N., Crow & Lounsbury 05189-2 7.50
Psychiatric/Mental Health Nursing Review,
    Second Edition, Rodgers & McGovern 04824-7 6.00

PROFESSIONAL CAREER EXAM SERIES

Action Guide for Executive Job Seekers and Employers,
    Uris 01787-2 3.95
Air Traffic Controller, Morrison 04593-0 10.00
The Anatomy of Arson, French .... LR 04423-3 12.50
Arson: A Handbook of Detection and
    Investigation, Battle & Weston .... LR 04532-9 9.95
Automobile Mechanic Certification Tests, Sharp 03809-8 6.00
Bar Exams 01124-6 5.00
Careers for the Community College Graduate,
    Chernow & Chernow 05091-8 5.95
Certificate In Data Processing
    Examination, Morrison 04922-7 12.00
The C.P.A. Exam: Accounting by the "Parallel Point"
    Method, Lipscomb .... LR 01103-3 25.00
Computer Programmer Analyst Trainee, Luftig 05310-0 8.00
Computers and Automation, Brown 01745-7 5.95
Computers and Data Processing Examinations:
    CDP/CCP/CLEP 04670-8 10.00
Dental Admission Test, Eighth ed.,
    Arco Editorial Board 05313-5 6.00
Graduate Management Admission Test 04914-6 6.95
Graduate Record Examination Aptitude Test 04910-3 6.95

Health Insurance Agent, Snouffer 04307-5 8.00
Health Profession Careers In Medicine's
    New Technology, Nassif 04436-5 5.95
How a Computer System Works, Brown & Workman ... 03424-6 5.95
How to Become a Successful Model—Second Edition,
    Krem 04508-6 2.95
How to Get into Medical and Dental School, revised edition,
    Shugar, Shugar, Bauman & Bauman 05112-4 6.95
How to Make Money in Music, Harris & Farrar 04089-0 5.95
How to Remember Anything, Markoff, Dubin & Carcel ... 03929-9 5.00
How to Write Successful Business Letters,
    Riebel 02290-6 5.00
The Installation and Servicing of Domestic
    Oil Burners, Mitchell & Mitchell 00437-1 10.00
Instrument Pilot Examination, Morrison 04592-2 9.95
Law School Admission Test, Candrilli & Slawsky 05153-1 6.95
Life Insurance Agent, Snouffer 04306-7 8.00
Medicine's New Technology, Nassif .... LR 04443-8 9.95
Miller Analogies Test 04990-1 5.00
Modern Police Service Encyclopedia, Salottolo 02389-9 8.00
National Career Directory, Gale & Gale 04510-8 5.95
The New Medical College Admission Test 04551-5 6.95

The Official 1981-82 Guide to Airline
  Careers, Morton 05238-4 **7.95**
The Official 1981-82 Guide to Steward
  and Stewardess Careers, Morton 05237-6 **7.95**
The Official 1981-82 Guide to Travel
  Agent and Travel Careers, Morton 05236-8 **7.95**
Notary Public 00180-1 **6.00**
Nursing School Entrance Examinations 01202-1 **6.00**
Playground and Recreation Director's Handbook ... 01096-7 **8.00**
Practice for U.S. Citizenship, Paz 05305-4 **2.95**
Preparacion para el Examen de la Licencia en Cosmetologia,
  McDonald & Mottram 05306-2 **6.95**
Preparation for Cosmetology Licensing
  Examination, McDonald & Mottram 04756-9 **6.95**
Preparation for Pesticide Certification
  Examinations, Frishman 04761-5 **10.00**
Principles of Data Processing, Morrison 04268-0 **7.50**
Property and Casualty Insurance
  Agent, Snouffer 04308-3 **8.00**
Psychology: A Graduate Review, Ozehosky & Polz ... 04136-6 **10.00**
The Real Estate Career Guide, Pivar 04790-9 **7.95**
Real Estate License Examinations, Martin 04794-1 **8.00**
Real Estate Mathematics Simplified, Shulman 04713-5 **5.00**
Refrigeration License Manual, Harfenist 02726-6 **12.00**
Resumes for Job Hunters, Shykind 03961-2 **5.00**
Resumes That Get Jobs, third edition,
  Resume Service 05210-4 **3.95**
Science Review for Medical College Admission,
  Morrison 04705-4 **12.95**
Simplify Legal Writing, Biskind 03801-2 **5.00**
Spanish for Nurses and Allied Health Science Students
  Hernandez-Miyares & Alba 04127-7 **10.00**
Stationary Engineer and Fireman 00070-8 **8.00**
Structural Design 04549-3 **10.00**
Successful Public Speaking, Hull .... LR 02395-3 **5.95**
The Test of English as a Foreign
  Language (TOEFL), Moreno, Babin & Cordes 04450-0 **8.95**
TOEFL Listening Comprehension Cassette 04667-8 **7.95**
Travel Agent and Tourism, Morrison 04746-1 **15.00**
Veterinary College Admissions 04147-1 **10.00**
Your Job: Where to Find It—How to Get It, Corwen ... 05131-0 **6.95**
Your Resume—Key to a Better Job, Corwen 03733-4 **4.00**

ADVANCED GRE SERIES

Biology: Advanced Test for G.R.E., Solomon 04310-5 **5.95**
Business: Advanced Test for the G.R.E., Berman,
  Malea & Yearwood 01599-3 **4.95**
Chemistry: Advanced Test for the G.R.E.,
  Weiss & Bozimo 01069-X **4.95**
Economics: Advanced Test for the G.R.E.,
  Morrison 04548-5 **5.95**
Education: Advanced Test for the G.R.E.,
  Arco Editorial Board 04714-3 **6.95**
French: Advanced Test for the G.R.E., Dethierry ... 01070-3 **5.95**
Geology: Advanced Test for the G.R.E., Dolgoff ... 01071-1 **3.95**
History: Advanced Test for the G.R.E.,
  Arco Editorial Board 04414-4 **5.95**
Literature: Advanced Test for the G.R.E. 01073-8 **3.95**
Mathematics: Advanced Test for the G.R.E.,
  Bramson 04264-8 **5.95**
Music: Advanced Test for the G.R.E., Murphy 01471-7 **3.95**
Philosophy: Advanced Test for the G.R.E., Steiner ... 01472-5 **4.95**
Physical Education: Advanced Test for the G.R.E.,
  Rubinger 01609-4 **3.95**
Physics: Advanced Test for the G.R.E., Bruenn ... 01074-6 **6.95**
Psychology: Advanced Test for the G.R.E., Morrison ... 04762-3 **4.95**
Sociology: Advanced Test for the G.R.E., Morrison ... 04547-7 **5.95**
Spanish: Advanced Test for the G.R.E., Jassey ... 01075-4 **3.95**
Speech: Advanced Test for the G.R.E., Graham ... 01526-8 **3.95**

GRADUATE FOREIGN LANGUAGE TESTS

Graduate School Foreign Language Test: French,
  Kretschmer 01461-X **4.95**
Graduate School Foreign Language Test: German,
  Goldberg 01460-1 **3.95**
Graduate School Foreign Language Test: Spanish,
  Hampares & Jassey 01874-7 **3.95**

PROFESSIONAL ENGINEER EXAMINATIONS

Chemical Engineering, Coren 01256-0 **8.00**
Civil Engineering Technician 04267-2 **10.00**
Electrical Engineering Technician 04149-8 **10.00**
Engineer in Training Examination (EIT), Morrison ... 04009-2 **10.00**
Engineering Fundamentals 04273-7 **10.00**
Fundamentals of Engineering, Home Study Program
  (3 Vols.) 04302-4 **45.00**
Fundamentals of Engineering (Vol. I), Morrison ... 04234-6 **17.50**
Fundamentals of Engineering (Vol. II), Morrison ... 04240-0 **17.50**
Fundamentals of Engineering (2 vols.) 04243-5 **35.00**
Industrial Engineering Technician 04154-4 **10.00**
Mechanical Engineering, State Board Examination
  Part B, Coren 01258-7 **12.00**
Mechanical Engineering Technician 04274-5 **10.00**
Principles and Practice of Electrical
  Engineering Examination, Morrison 04031-9 **10.00**
Professional Engineer (Civil) State Board
  Examination Review, Packer et al 03637-0 **15.00**
Professional Engineering Registration: Problems
  and Solutions 04269-9 **10.00**
Solid Mechanics, Morrison 04409-8 **10.00**

NATIONAL TEACHER AREA EXAMS

Early Childhood Education: Teaching Area Exam
  for the National Teacher Examination 01637-X **6.95**
Education in the Elementary Schools: Teaching Area
  Exam for the National Teacher Examination 01318-4 **8.00**
English Language and Literature: Teaching Area
  Exam for the National Teacher Examination 01319-2 **3.95**
Mathematics: Teaching Area Exam for the
  National Teacher Examination 01639-6 **6.00**
National Teacher Examination 00823-7 **6.95**

TEACHER LICENSE TEST SERIES

Guidance Counselor—Elementary, Jr. H.S. & H.S. ... 01207-2 **7.00**
Teacher of Common Branches 00770-2 **6.00**
Teacher of Early Childhood, Elementary Schools—
  Kindergarten to Grade 2 .... 00771-0 **6.00**
Teacher of English, Jr. H.S. & H.S. 00790-7 **8.00**
Teacher of English as a Second Language, Wellman ... 04024-6 **8.00**
Teacher of Fine Arts, Jr. H.S. & H.S. 01037-1 **6.00**
Teacher of Industrial Arts, Jr. H.S. & H.S. 01307-9 **6.00**
Teacher of Mathematics, Jr. H.S. & H.S. 00816-4 **8.00**
Teacher of Spanish, Jr. H.S. & H.S. 01027-4 **7.00**

TEACHER PLAN BOOKS

Arco Teacher's Plan Book—Elementary School
  (Grades 1-8) 04029-7 **3.00**
Arco Teacher's Plan Book—Jr. & Sr. High School
  (Grades 7-12) 04028-9 **3.50**
Arco Visible Record Books (Looseleaf 6 x 8) 01281-1 **6.00**
Arco Visible Record Books (Looseleaf 8 x 8) 01280-3 **6.25**
Cards for Arco Visible Record Books 00761-3 **3.00**
Leaves for Arco Visible Record Books (8 x 8) ... 07901-0 **.85**
Leaves for Arco Visible Record Books (6 x 8) ... 07900-2 **.85**